SPOILER ALERT: EVERYONE DIES™

A Whimsical Journey through the Lighter Side of Global Annihilation

SPOILER ALERT: EVERYONE DIES™

A Whimsical Journey through the Lighter Side of Global Annihilation

DAVID CONSIGLIO, JR.
With Eddie Wetmore

B E G Publishing, LLC

SPOILER ALERT:
EVERYONE DIES™

Bell MT fonts used in varying sizes.

Library of Congress Control Number: 2017943283

CreateSpace Independent Publishing Platform, North Charleston, SC
Although intended to give honest answers to hypothetical questions, this is a work of fiction. Any similarity between the characters and situations within its pages and places or persons, living or dead, is unintentional and coincidental.
ISBN-13: 978-1542893022
ISBN-10: 154289302X
BISAC Category: SCIENCE/PHYSICS/GENERAL
First Edition 2017
Cover design and layout by Consiglio Creative

Acknowledgments

A ton of people helped write either comments of inspiration for this work-including; but, in no way limited to, Josh Velson (fantastic treatise on Earth's conversion to Heavy Water), Kim Wetmore (for several unhelpful comments concerning employability when I was playing computer games instead of working on the computer, like I was supposed to) and Amy Consiglio, for some brutal grammatical editing that kept us from sounding like blithering idiots, as well as, designing a cover that—quite literally—blew us ALL away!

CONTENTS

"Then you should say what you mean," the March Hare went on.

*"I do," Alice hastily replied; "at least—at least I mean what I say—
that's the same thing, you know."*

*"Not the same thing a bit!" said the Hatter. "Why, you might just as well say
that 'I see what I eat' is the same thing as 'I eat what I see'!"*

Excerpt from chapter VII, "A Mad Tea-party,"
from Lewis Carroll's Alice in Wonderland:

Foreword by Dave Consiglio

It all started with a simple question: **"Assuming we tie an unbreakable rope to the moon, can the Earth's population pull it to the Earth?"**

The question had been answered by perhaps a dozen people already, and they all said the same thing:

"It can't be done."

They're right, of course—for countless reasons this is impossible. But those answers upset me. They answer the letter of the question, but they speak no words to its spirit. The question may seem silly on its face, but how often have you felt silly about not knowing a thing?

I thought the answer should take that into account and be silly, too.

So, I decided to throw out the major reasons it couldn't be done, and approach the problem in a lighthearted way.

The seed of Everyone Dies™ was born.

Since then, I've tried to look into the heart of ridiculous questions and pull out a kernel of satisfying truth. Sure, we can't really make a rope to reach the Moon. Sure, even if we could, the rotation of the Earth would pull this rope away from us at preposterous speeds.

But, what of it?

The learning, and the humor, is found beneath the impossibilities.

This book is, therefore, dedicated to those brave enough to ask silly questions, and those willing to answer them. I hope this work strips away a little fear from everyone who ever felt stupid about not knowing a thing, and a little arrogance from those who show disdain toward those who know less. We all begin life full of ignorance and curiosity—this is the place from which learning really happens.

Embrace your ignorance, and never stop seeking to fill the holes in your understanding with knowledge and wit.

I'd like to thank the Quora community for providing me with an unending stream of questions, comments, and support. To the team at Consiglio Devastations—your work has been an inspiration to me, and I will be forever grateful for all that you've done. Your passion and dedication to the cause of education through the wanton destruction of planets is a joy in my heart.

In particular, I'd like to thank three very special people: Dr. Habib Fanny, Amy Consiglio, and Eddie Wetmore.

Habib—you're an inspiration in the real world, and for that I'm truly thankful. I would never have gotten on Quora, were it not for you. But, more importantly, America, and humanity at large, needs stories like yours to remind us of the fact, all too easily ignored by so many, that people who seem fundamentally different from us are exactly the kind of people we should embrace and welcome into our communities.

Diversity in all its forms really is a strength.

Amy—you're an inspiration to me every day. Your support of my crazy endeavors is truly superhuman. You have always thought more of my abilities than I have, and your confidence has given me the courage to branch out and do more. Thank you so much for your patience and understanding, as well as your amazing editing skills! A lesser woman would have written this off as a stupid hobby—I'm thankful every day that I married someone better than I am, in all the ways that count.

And Eddie—what can I say about this guy? His dedication and devotion to this project astonishes me. His humor and personality drip off of every word he writes. I don't know Eddie nearly as well as I would like to, but I would never have met him at all were it not for this book. You don't know this, but I lost a special Eddie a few years back. You remind me a little of him. So thanks, Eddie. You're two in a million. What a truly lucky man I am to have known you both. I look forward to many heavy conversations and projects in the future.

To all the readers—I hope you enjoy what follows.

Keep asking silly questions and promoting understanding wherever you go.

Everyone Dies™ may be my trademark, but my dream is that Everyone Learns™.

Dave Consiglio
February 5, 2017

And let the Destruction Begin!

Assuming we tie an unbreakable rope to the moon, can the Earth's population pull it to the Earth?

Ooh! This is such a fun question.

OK, let's forget all about the rotational nightmare this introduces (we're running at around 1,000 miles per hour to keep up with the rope at the equator). In addition, there's essentially no friction in space, so we could pull any object into the Earth, given enough time.

Instead, let's look at slight variation of your question: If we stopped rotation and revolution and held the Earth and the moon still, ignored gravity*, and also somehow managed to get 7 billion people on the same rope (or each holding a connected strand), how long would it take to pull the moon down if we all pulled at the same time?

The average pull strength of a human is a hard thing to estimate, but for adults with experience, it is in the 100 - 200 pound (45-90 kgs) range. But these are burly canyoneers! What about all of the babies and older folks that aren't quite so brawny?

I'll use an average of 50 pounds (23 kgs) for humans. Sorry, babies, you're just going to have to try your best. Now, a pound of force is about 4.5N (let's get out of this ridiculous American affection for old-fashioned units). There are about 7.2 billion humans, which gives a force of 1.62 trillion Newtons.

The mass of the moon is 7.35×10^{22} kg. Thanks to the aforementioned Newton, we know that $F = ma$. The acceleration we produce on the moon is F/m, which is 2.2×10^{-11} m/s^2. That's pretty small! The distance from Earth to moon is about 370.3 million meters. (3.703×10^8).

So (thanks again, Newton) since $d = \frac{1}{2}at^2$, solving for t, we get which equals almost exactly 184 YEARS.

When the moon gets to us, it's traveling at a paltry 0.127 m/s, or about 0.25 mph. That's really slow. When it hits the Earth, though, it has a kinetic energy of about 5.9×10^{20} J, or about 141,000 megatons of TNT. This is many times more than the total of all nuclear testing ever done, and would probably wipe out life on Earth as the moon slowly burrowed hundreds of miles into the mantle.

*OK, we ignored gravity. Suppose we skip the rope and just let Earth do the pulling for us?

This is the force Earth exerts on the moon (and the force the moon exerts on the Earth). In this case it's equal to 2.13×10^{20} N, or about 10 million times stronger than our pitiful little band of rope pullers.

If we include that force, Earth and moon would collide in less than 6 days and the impact would be billions and billions of times more powerful than the previous impact—the moon would be traveling at more than 3,000 mph. This is more than enough to blow both the Earth and the moon to tiny bits.

Everyone Dies™.

(Editor's note: And thus the legend is born...)

Which Everyone Dies™ has the greatest variety of death?

An unstable star that has many Coronal Mass Ejections. Consiglio Devastations gets to work on this record-setting devastation.

1. The first CME hits Earth and starts to ablate the atmosphere of our planet, killing a few people instantly (like on the ISS). (1 - irradiation)

2. Cancer rates skyrocket, killing many more. (2 - cancer)

3. Humans decide to leave the solar system, thanks to the instability of our sun. We spend a century hollowing out an asteroid for our first interstellar trip to Proxima b.

4. In the meantime, society starts to crumble, leading to riots (3 - trampling), murders (4 - blunt force trauma / bullet wounds), and the theft of medicine and food (5 - diabetic shock, 6 - starvation). Poor sanitation leads to cholera (7 - waterborne illness) and the accumulation of pollutants (8 - heavy metal poisoning).

5. Nuclear power plants fall into disrepair, and a few go critical. These explosions, coupled with an accidental nuclear strike, give me another death cause (9 - vaporization).

6. As the CMEs increase in frequency, temperatures on our planet soar (10 - heat stroke) and UV protection dwindles (11 - sun poisoning). In addition, mutations begin to crop up everywhere (12 - genetic diseases).

7. The last humans on Earth are kept safely underground until it is time to launch them on Salvation 1, our first (and last) interstellar ship. There is a malfunction on several of the rockets that take people there, resulting in (13 - asphyxiation) and (14 - burning).

8. The few remaining humans make the trip to Proxima b, though several are killed due to the rough nature of the landing (15 - traumatic aortic rupture due to rapid deceleration).

9. Sadly, the last few humans don't last long - they are eaten by the natives (16 - becoming food).

Everyone Dies™.

Or, as Paul Simon might put it:

There must be 16 ways to end the humans…

Slip on a crack, Jack
Burn off your skin, Quinn
Don't need to be fed, Fred
Just listen to me

Fall off a bus, Gus
Don't need to discuss much
Stop being alive, Clive
And Everyone Dies™

Valentine's Day is coming up. How do I get my girlfriend something nice and also ensure that Everyone Dies™?

Ah, love. Is there anything more precious than that first glance, the warmth in your face after your first passionate kiss, and the glow in her eyes from the radiation emitted from the accretion disk you got her for Valentine's Day?

No, no there is not.

So, being the connoisseur of love that I am, I'm going to try and make this Valentine's Day extra special for you and that special lady in your life. She deserves it, and so do you.

First, sweep her off her feet with a hypersonic air blast created by a thermonuclear weapon specially designed to create a mushroom cloud in the shape of a heart, burning with your love for her.

Then, when she's still breathless (from the impact), give her that tingling sensation as she basks in the glow of the tropical sun. Don't worry if you can't afford that trip to Tahiti - you can take her to Duluth, instead. It's remarkably inexpensive, and a lot warmer now thanks to the orbital manipulations that have brought Earth dangerously close to its sun. Sure, you've increased her risk for melanoma by several orders of magnitude, but for a little while, it's just a warm glow.

After a glorious meal of the last steak on Earth, it's time to turn up the passion. Watch her inhibitions melt away, along with the rest of her, as you point to the heavens and show her the gorgeous swirl of the moon being ripped to atoms by the black hole you've parked right off its surface. The black hole has been made to spin so that the last thing you and your beloved see as your bodies are ablated by horrible radiation is a disk that looks remarkably like a dozen roses.

I promise that you will spend the rest of your lives together!

Happy Valentine's Day from Consiglio Devastations.

Everyone Dies™.

What would happen if the Moon was briefly replaced by a full-spectrum electromagnetic emitter with a power output of 10^{31} Watts?

For 10 seconds, the Moon was replaced by an object emitting 10^{31} watts worth of the entire electromagnetic spectrum (from radio waves to gamma rays) in every direction.

You're making it too easy for me!

The output of our sun is 3.846×10^{36} W. So your emitter is around 26,000 times more powerful than the sun, and is 400 times closer.

But we're only getting this new Super Sun for 10 seconds. During that time we're going to get 1×10^{32} J—this is around three days worth of solar radiation in just 10 seconds, all on one side of the Earth.

So what happens?

The side of the Earth facing this emitter bursts into flames. Forests, houses, even people spontaneously combust. Much of the polar ice melts. Sandy deserts get covered with a thin layer of glass.

Oh, and the air itself burns. All of that radiation creates ozone (O_3) from atmospheric oxygen (O_2). Ozone reacts with nitrogen, forming NO_x, a horribly poisonous combination of NO, N_2O and NO_2, along with lesser amounts of more complex nitrogen oxides.

The other side of the Earth fares only slightly better. For starters, between all of the burning biomass and the reactions in the air, there is probably not enough oxygen left to breathe. In addition, all of that combustion on the other side of the planet is going to throw absurd amounts of soot into the air, blocking out the sun and throwing Earth into years of perpetual darkness and cold. But first, the rapidly melting ice creates horribly unstable ice caps. As enormous chunks collapse into the sea, giant tsunamis circle the globe, devastating coastal cities. The shifting mass triggers earthquakes, too.

I probably don't even need to say it, but it's very unlikely that anyone survives this. The reaction of all of that atmospheric oxygen most likely asphyxiates humanity (and most other animal life). Even after the emitter is gone, bacteria will rot what didn't burn, consuming even more oxygen. Nearly all plants will die due to the global darkness—they'll rot, too, consuming even more oxygen. And those dead plants are not producing new oxygen to replace what is lost.

So the world burns, then chokes, then freezes.

Everyone Dies™.

How fast would someone have to be running to kill another person on impact?

Please only consider the energy of just running into another person, full-speed, and causing their death on impact. You can assume that other person is their identical twin if it makes physics/math work out well.

Oh, I just started salivating when I read this one!

First, some caveats:

1. Killing another person is a highly variable thing. The same impact with the same exact force delivered to various locations might or might not kill someone. Hit someone in the leg with a car, and they might survive. Hit them in the chest, and they might die.

2. The "softness" of the person doing the impacting complicates matters. When you're hit by a car, the car crumples, but not that much. The victim, on the other hand, tends to crumple a great deal. But humans aren't uniformly crumply (a sentence I am so glad I have gotten to write...) If I hit you with my shoulder, I'll do more damage than if I hit you with my rear end.

3. The environment makes a big difference here, too. Are you landing on concrete? Grass? Are we ignoring the ground and just focusing on the impact?

So, let's look at the rough approximation of hitting a person with as much energy as a car, and ignore the ground.

The average car has a mass of around 1500 kg. If you get hit at say 50 kph, you're probably likely to die. A car at this speed (around 14 m/s) has an energy of 147,000 J.

Now, your average man has a mass of around 70kg—women are a bit lighter on average. In order to have the same amount of kinetic energy as the car, a 70-kg man will have to run at around 64.8 m/s. That's around 145 mph (233 kph). This is, of course, far faster than any human can travel (without help), but to me it feels a bit high.

My guess is that a human wouldn't need quite as much kinetic energy as a car because much of the car's kinetic energy wouldn't be transferred to a person hit by that car. If you're hit by a car, you most likely fly into the air or get run over. In both of these situations, the car continues moving, not transferring all of its energy to you. Instead, there's extra force involved: gravity. The flight through the air results in a massive impact on the ground, compounding the injury. Similarly, being run over by the car adds crush injuries to the list of problems you're likely to face.

But, a person traveling at 145 mph would most likely burst you like a melon if he hit you head on. Chances are a much lower speed would do the trick.

There's historical precedence for this (believe it or not): Harry Houdini.

Harry Houdini may have died due to several punches in the stomach. Houdini was famous for his many "magical abilities", including the ability to be punched without suffering injury. But Houdini was lying down at the time he was last punched, and the punch may have injured or ruptured his appendix, leading to his death.

No matter who punched him, the total energy was far less than 147,000 J.

A punch has an energy of around 100—450 Joules, or perhaps 1,000 times less than the car. But, that force is concentrated in a small area.

However, if a person could run and concentrate the energy using their shoulder, the point of contact wouldn't be much larger than a fist.

And, to get that much energy, a 70 kg man would only have to run at the incredibly low speed of 3.6 m/s, or around eight miles per hour…a speed most people can run at (albeit briefly).

This number, however, seems too low—if this were true, the NFL would probably be riddled with deaths from direct shoulder-to-stomach impacts.

So, the answer is almost certainly less than 100mph, but probably higher than 8mph.

The last bit of information I could add is that death from falling from a 3rd story window is relatively common. If that window is 30 feet above ground, then the falling person has a speed of around 14 m/s (31 mph). That's not quite the same thing as being run into, but, it's reasonably close.

My final answer is around 30 mph. Incredibly, Usain Bolt can run almost this fast.

So, if you get a chance to go to the Olympics to see Mr. Bolt run, do not get in front of him. He might just be running fast enough to kill you.

Is Earth a dying planet?

My friend asked this question and I did not know how to quite answer it. The question was simple, is the Earth flourishing or dying?

All planets are dying.

The minute a planet is formed around a star, its days are numbered. That star will burn through its hydrogen fuel in the ordinary way, taking more or less time depending on its mass. But sooner or later it will run out of its fuel and the planet will freeze solid, killing any processes that once made it an interesting place.

But then flourishing and dying are not mutually exclusive. The Earth is flourishing and dying, and so are you and I. I am flourishing while I write this answer, just as sure as I am shorter of breath, one day closer to death.

Tell your friend it's not a good question—the Earth is flourishing and it is dying.

How dangerous would guns be if they used antimatter as an accelerant?

Just assume the gun doesn't explode, and the recoil stays the same somehow.

And assume the bullets and gun are built in a way to that they actually work with antimatter.

The bullets are still kinetic and don't explode.

The amount of antimatter is the same as the amount of the explosive stuff used before.

0.5g of antimatter reacts with 0.5g of matter to produce around 90 trillion Joules of energy. But that also means that 0.5 nanograms of antimatter reacts with 0.5 nanograms of matter to produce around 9 thousand Joules of energy. That's a pretty average muzzle energy for a gun.

So, an amount of antimatter too small to see could easily kill someone. This could allow for a practically unlimited amount of ammunition—1.0 g of antimatter could be used to turn two billion bullet-shaped lumps of metal into deadly projectiles.

But, you wouldn't even need bullets—if you could focus the radiation, you could just aim and shoot electromagnetic death at your victims. The proper radiation dose would make your victims cook almost instantly, similar to what beam weapons in science fiction are capable of doing. Lower amounts of antimatter would just give your victims radiation poisoning or cancer.

It's a good thing antimatter is extraordinarily difficult to make. A gram could turn its owner into the world's most devastating terrorist, and a bathtub full might shatter the Earth!

What would happen if a second Moon appeared exactly opposite the Moon that already exists?

Effectively a duplicate Moon, same size and mass, on the same orbit, just delayed by 1/2 an orbit cycle.

You know, sometimes I feel like I exist here just to tell people they're going to kill everyone on Earth. So…ahem….

We're going to kill everyone on Earth. (!)

(Editor's Note: Refer to: What would it take to move Venus closer to Earth, making Earth/Venus a twin planet? Page 57)

Putting two bodies on opposite sides of a larger body is an unstable configuration. Tiny variations caused by the gravity of the Earth, sun, and other planets will speed or slow one of our two Moons. This will get the two Moons closer to each other, and their own gravity will take over from there. In very short order, the two Moons will collide. This will send enormous chunks of Moon rock hurtling in all directions…including toward Earth. The impact of something that large wipes out life on our planet and liquefies the crust.

It would be really pretty, followed by explosive, followed by really ugly. We would actually be able to watch the impactor hurtling toward our world. As we rotated, everyone would get to see their death plummeting down to Earth in the form of an irregular, glowing hunk of destruction. Within mere days, it would be nearly on top of us, building up speed as it fell, moving fast enough to actually see it moving from the Earth.

I can see it now, laying on my roof with my family, noshing on popcorn, explaining to my young ones how some 'blockhead' had this crazy idea to add another Moon. "Seemed like a good idea at the time," I'd say, and chuckle. Then, I'd point out the faint orange streak behind the hunk of Moon we'd been watching. "Do you see that, kids? That's ionized rock glowing from the heat of reentry. It won't be long now. In just a few moments, you'll see the".

And then the shockwave would streak across the planet, demolishing everything in its wake. Our family would be vaporized, leaving behind only the faintest imprint in the bedrock as evidence that our home once sat here.

So yeah, Everyone Dies™.

Millions of years hence, alien astronauts would arrive here and see a hunk of rock where once a teeming planet existed. "G*@OI" they would say, which of course translates into, "That's odd. A planet at this distance from such a star should be teeming with life. Such a shame. They must have been struck by a massive impactor…"

If Jupiter is a failed star that never started fusion, what would the solar system be like if it had?

To become a star, Jupiter would need more mass.

The asteroid belt is an unstable region due to the interplay of gravity between the sun and Jupiter.

A more massive Jupiter would move that band of instability in toward the sun—and toward Earth.

The increased radiation from another star would also wreak havoc with planetary formation.

Chances are there would be no Earth at all, and probably no "inner planets". Outer planets would have to orbit far further from both stars in order to prevent their orbits from being perturbed and then being ejected from the solar system. That means Saturn couldn't exist, and maybe not Uranus or Neptune.

In short, our solar system would be totally different.

What would happen if we split Earth's atmosphere between the Earth and Moon?

Judging by surface area and gravity, these evenly split atmospheres would have an equilibrium pressure of ~0.684 atm. I ruled that Earth would remain habitable—if less comfortable—but what about the Moon? Would the surface oxidize and be affected by aeolian processes, and how long would it last?

The Moon would quickly lose the air we gave it and we would all die.

The Moon doesn't have enough gravity to hold on to gases at such high temperatures. It also completely lacks a magnetic field. These two factors would allow the Moon's atmosphere to leak into space in short order, rendering it quickly uninhabitable.

As for Earth, I doubt we would survive this. 0.684 atm of pressure is equivalent to an altitude of over 10,000 feet. Human beings often suffer from altitude sickness at just 8,000 feet. So if you quickly split our atmosphere, many human beings would suffer from hypoxia and die. Furthermore, many animals are less capable of surviving in oxygen-poor environments than we are, as are many plants. These, too, would also perish.

Some humans would survive, but they would have a major problem on their hands. First, the bodies of billions of creatures around the world would begin to rot, consuming even more oxygen. Our oceans would start to rapidly evaporate, significantly increasing the amount of water vapor in our atmosphere. This would increase global warming and bake our planet. That's bad news for life, too, and might make our planet more like Venus in the long run.

So, even if some people survived the initial loss of oxygen, they would continue to lose oxygen rapidly as so many dead creatures rotted. The temperature would first plummet, and then slowly rise to far warmer than it is today.

Everyone Dies™.

Will a person survive longer in Jupiter's upper atmosphere (one bar of atmospheric pressure) than in outer space?

Given a person is completely unprotected by a spacesuit, will s/he survive longer in upper layers of Jupiter's atmosphere than roughly ninety seconds? Can a person breathe in a gas in the upper layer of Jupiter? What will happen to a person in that environment? Upper layer, One bar of atmospheric pressure.

Yes…if the person holds his or her breath.

In empty space, you've got about 10 seconds of consciousness, maybe 15 seconds of confusion, then you black out. You're dead in two to four minutes. You can't physically hold your breath—the pressure of the air in your lungs is too much for your muscles and the air escapes.

At 1 bar in Jupiter's atmosphere, you could hold your breath. This means even an average person could stay conscious for over a minute and not die for maybe five to ten minutes.

But, if you breathe, the tables are turned. You're getting a lungful of ammonia and carbon monoxide and hydrogen cyanide and all sorts of other things that will expedite your demise.

So, hold your breath, and you'll last a bit longer on Jupiter. Breathe, and you're a goner on Jupiter a little quicker.

But, it's really pedantic—in both cases you're dead in minutes.

How would one destroy Earth on his own?

Earth is big. REALLY big. Destroying it in the traditional sense (making it no longer exist) is outside the realm of the possible for humanity, much less one person...unless you get a little help from the universe, of course.

It might be possible to steer a large asteroid into a collision course with the Earth. This requires a large asteroid to pass very near the Earth, of course. This would still take the resources of a space agency, but if there was a mission planned to study the asteroid, it might be possible to sabotage it in such a way as to divert the asteroid into a future collision with Earth.

But, short of that unlikely chain of coincidences, you're stuck with the Earth for a long time to come.

How dangerous would a gun be if bullets traveled with 99.9% the speed of light?

It depends on where you fire it.

If you fired it in space, the bullet would go right through you, just like an ordinary bullet. But, as it passed, it would press the hydrogen atoms in your body together and cause them to fuse. This would create a tiny fusion explosion in the path of the bullet, and by tiny, I mean fusion tiny, which is huge. Your body would be obliterated in a rain of radiation.

If you fired this gun on Earth, the bullet would also cause fusion to take place, but this time in the air. The energy creates a ball of plasma and obliterates the shooter and the victim, regardless of whether the bullet hits him or not.

What if the Earth and Mars merged into one planet?

We would get one planet that is about 10% heavier and only slightly larger than the Earth. My guess is that life could easily evolve on such a planet, as it did on Earth, and that the chemistry and biology of that life would probably be similar to what evolved here on Earth.

Of course, actually bringing Mars and Earth together would sterilize both planets, turning them into a seething ball of molten rock and metal before the wreckage cooled into a place where life might be able to survive. This would take millions of years.

And, of course, Everyone Dies™.

If a gamma-ray burst (GRB) was headed our way, how will it affect our planet?

TOTALLY an imaginative and extremely unlikely scenario, for sure, but hey let's just say that it did happen. How exactly will our planet be affected? Would we see it coming? What about other planets and the Sun? What will happen to them too? Thanks!

Like all emissions of energy, the poison is in the distance.

We get hit by GRBs all the time—that's the only way we can see them. But since most of them take place millions or billions of light years away, it takes powerful telescopes to spot them at all. The resulting energy imparted to the Earth is so small that the average human is incapable of perceiving it.

If, on the other hand, a GRB took place say on the Moon, it would literally rip planet Earth apart and blow the debris clear out of the solar system.

So, the answer ranges from "Meh" to "OMG NO!" depending on your distance.

How hot would the world need to be to evaporate 70% of the world's oceans.

What would the world look like, and what would the terrain be like? Where would the water go?

The water would go into the atmosphere. That part is easy. But how hot is hard. The problem is that as we increase the temperature, the mass, and thus pressure, of the atmosphere will also increase. This will change the equilibrium conditions of the system. In addition, we'll change the solubility of gases in the oceans, the amount of CO_2 in the atmosphere from decomposing carbonate rocks, and the amount of SO_2 in the atmosphere from decomposing sulfite and sulfate rocks. In short, this is a complicated problem.

The mass of the atmosphere is around 5×10^{18} kg. The mass of the oceans is around 1.4×10^{21} kg. So boiling off 70% of the world's ocean will add around 1×10^{21} kg to the atmosphere, increasing its mass by a factor of around 200. This will increase the pressure by an even greater factor. Water has a lower molar mass than air—so pound for pound you get more molecules, and thus more pressure, from water than from air. This has the effect of turning Earth into an absolute pressure cooker.

It turns out that increasing the pressure this much puts us very close to something special: the critical pressure. At 218 atm and 374° C, the distinction between water vapor and water liquid ceases to exist, and instead you have supercritical water.

I don't think there exists a temperature where 70% of the water on Earth has boiled off. I think if you got it that hot, the pressure would increase so much that you'd exceed the critical pressure and lose the distinction between liquid and gas. But in all fairness, I'm not terribly confident about this answer. It's a very complex question you've asked!

However, I'm pretty sure that (drum roll, please)...

Everyone Dies™.

If suddenly humans became unable to reproduce, in what year do you think humanity would become extinct?

When would the last human die assuming people currently pregnant will give birth to the last generation of humans?

Just over 90 years from now.

The infants on the planet could live past 100...with the help of society. But, by the time they're 90, there will be no society. Everyone else will be dead. Without doctors and modern medicine, not to mention grocery stores and electricity, the aged won't do as well. A few unusually spry oldsters could hang on, gardening and scraping by on the remains of civilization. But sooner or later they're doomed anyway, so most would probably give up.

What a dreary picture.

What would Earth be like with zero oxygen levels?

Gasp…gasp…ack…thump.

(editor's note: Bwahahahahaha! Gotta give points for succinctness…)

If there was no oxygen in the atmosphere, almost all animals would die. Plants would die also, as they need oxygen for respiration. Some extremophiles might survive, but it'd be pretty dire here on Earth.

If there was no oxygen on Earth, not only would we have no breathable air, but we would have no water and almost no rock. We would be a giant ball of iron and nickel with some silicon and sulfur and a few other elements thrown in for good measure. In that regard, we'd be much more like Mercury.

It goes without saying Everyone Dies™.

Physically and environmentally, what would happen if the US exploded all its 7,000 nuclear warheads at once on The Marquesas Islands?

The Marquesas are the farthest from any continent of all islands in the world. Let's assume that variable warheads are set to maximum yield.

These weapons would pulverize the Marquesas Islands and throw them into the air. This would cause major climatic changes.

In addition, the radioactive fallout would blanket the planet, increasing the rates of cancer and radiation poisoning. Ocean life would be particularly hard hit, with the highest radiation poisoning surrounding the former location of the islands we just eradicated.

Would it be nuclear winter? Maybe, maybe not. There's a lot of debate in this area. But I think there is no question that climate would be dramatically altered for the worse.

A year without summer? Almost certainly.

Years without summer? Possibly-might even be likely.

Snow in July in the breadbasket of America? Maybe.

The collapse of society? Perhaps.

No matter what, though, this is categorically a bad idea.

Suppose you had near-infinite energy and you could fire Deimos at near the speed of light aimed straight at Earth's core. What would happen?

Oh, Deimos would live up to its name!

Even ignoring relativistic effects, at .99c, Deimos would have almost 10^{36} J of energy. This is four orders of magnitude higher than the gravitational binding energy of Earth.

You know what that means???

Earth is not going to survive this encounter.

Deimos will make the trip from Mars orbit to Earth in mere minutes and it will blow through the Earth in less than a second.

But, during that fraction of a second, it will impart much of its kinetic energy into the Earth. This energy rips the Earth into tiny pieces which fly forward and out at significant fractions of the speed of light. Most of these pieces have more than enough velocity to leave the solar system—they are destined to roam the galaxy until they are pulled into other star systems.

Some of the debris pelts the Moon, obliterating it as well. More of the Moon stays in our solar system, along with tiny bits of the Earth. These pieces coalesce into a new, small planet.

And millions of years hence, alien explorers wonder why the 3rd planet from a nondescript star looks so young…

Everyone Dies™.

What would happen if the Moon was teleported to the surface of the Earth in the middle of the ocean?

The Moon on the surface of the Earth without a major impact, could it crumble into a large mountain or break through the crust from its weight? How much would the water level rise, could it just look like a big white ball in the middle of the ocean?

The Moon and the Earth are gravitationally attracted to one another. That attraction varies with distance. The closer they are, the stronger the force.

The force between them would be so strong in your scenario that Earth would pull the Moon right through the crust, right through the mantle, and into the core.

This would liquefy the crust, melting all of the continents and boiling off the oceans.

Earth would be significantly hotter and now, thanks to an atmosphere rich in water vapor, a runaway greenhouse effect would bake our planet like Venus. We would never be as hot, but temperatures of several hundred degrees would be the norm for many years to come.

I love the smell of catastrophe in the morning. Smells like victory.

Everyone Dies™.

Hypothetically, how long would the Earth survive in the absence of frictional forces?

Without friction your body, your house, and your planet would fall to pieces in mere milliseconds. Earth would collapse without the friction provided by electron expulsion and you, me, and everything else would contract under gravity into a singularity. Every other massive body in the cosmos would do the same.

Thanks—you've destroyed the entire universe. That's a first for me!

What if Earth were the biggest planet?

I read this two ways:

1. What if all planets were smaller than Earth?

2. What if Earth was as big as the largest planet?

If all planets were smaller than Earth, we would probably be dead. Jupiter keeps large space rocks from turning Earth into a scorched smoking crater land. Saturn and Uranus and Neptune do their parts, too.

If Earth was as big as the biggest planet, we would be living on an enormous gas giant. Briefly. Then we'd fall deep into the hellish inferno of the interior.

For the first time I get to say it TWICE: Everyone Dies™ or Everyone Dies™.

Every person on Earth has one day to live. Killing another person will give you their day in addition to yours. How long will the human race last?

Around 5,000 - 10,000 years—for just one very special man.

Humans are **HARD** to kill. Most people don't have the means to kill many people, and what's more, they most likely wouldn't want to. Would *you* kill someone just to stay alive for a day? I wouldn't. I wouldn't want to be alive in that world.

So, after the first day, billions are dead. But there are a few people with the means to kill *lots* of people. I don't think any of them would use that power, except for one: Kim Jong-un.

Suppose the dear leader used nuclear weapons to kill many people in China. He could easily buy himself a normal lifespan—only 14,600 deaths will get him into his 70s.

But, one nuke on a large city kills *millions* of people—and this could make Kim nearly immortal. If Kim nukes three million people, he nets himself around 8,100 years of life.

But, just a few days into Kim's near immortality, almost no one is left—and what's worse, most that remain have just a few days to live.

After a week or two, the only people left are mass murderers. It might be *just* Kim.

So, the dear leader gets to spend his thousands of years incapable of dying because there is no one left to kill him. He cannot procreate to keep this going, by the way—if he had tried to impregnate thousands of women, they wouldn't be able to live nine months to deliver the babies. Even Kim can't fight math.

Kim has become the supreme ruler of Earth—it was his lifelong dream. Sadly, it has become his genuine lifelong nightmare.

And around the year 10,000, while out for his daily walk, Kim stumbles to the ground and dies.

He was the last man on Earth.

If a black hole is hurtling towards Earth and we have 120 years until impact, what can we do that is theoretically possible for us to survive?

We're gonna need an asteroid. Stat.

We cannot deflect a black hole. I'm not sure we can deflect a tiny meteor! We need to get out of this vicinity and we're gonna need a big spaceship to do it. Why build one where there are lots of spaceships waiting to be colonized?

So, we take our asteroid, we hollow it out, and we colonize the inside. This takes 100 years but that's OK. We have the time.

As the black hole gets close, we use fusion or antimatter power (or maybe solar sails) to get into position so that we can use the black hole to slingshot us to another solar system. Might as well use our destructor to help us.

In a few decades or centuries we arrive at the exoplanet of our choice.

Our colonists set up a brave new society on a virgin planet.

Earth is destroyed, but humanity lives on.

What is the smallest thing in the human body that could be destroyed/removed and completely disable/kill the person?

One small group of atoms will do the trick. Something a few dozen nanometers across. A nucleotide base in the DNA of just one cell.

Chosen properly, this single omission could lead to a fatal cancer. It would take a while, but it would meet your criteria. It would be so small you couldn't see it. Even the most powerful microscopes on Earth would be able to resolve it only as a tiny patch of dots.

What if Earth suddenly became six times its volume?

Earth's density is just under six, so increasing our volume by a factor of six would get our density under one. We would be less dense than water!

The question would be how you did it. If you filled the Earth with pockets of empty space, the Earth would just collapse under its own gravity. While you would kill everyone on Earth, our planet would return to its initial size, mass, and density.

If you increased the volume of Earth by changing the material from which it is made, you'd have different problems. If Earth were made from water, its density would be about 1/6th what it is now. But, its significantly reduced gravity would not be enough to hold down a thick atmosphere and it would begin to freeze. Earth would be more massive than Mars, but only barely. This suggests a thin atmosphere with a temperature significantly colder than what we have now, but warmer than Mars. Earth's ocean might have portions that were liquid, but in the long term, I suspect Earth's ocean would freeze over. This would decrease the thickness of our atmosphere and increase our albedo, both of which would lower our temperature.

Either way, Everyone Dies™.

How can we destroy the Moon?

We can't. I'm done. Seriously, this can't be done...

OK, OK, it *can* be done. But, understand this is no small task. The Moon is really big. So, let's define some terms, first:

1. Destroy: This is pretty vague. Do you mean just break it up into chunks? Vaporize? Remove from the solar system? Let's go with "YES!" and do all of those things.

2. We: By "we", I'm assuming you mean humanity, as it currently stands. This stops the question in its tracks. "We" cannot destroy the Moon, or come anywhere close to destroying the Moon. All of the rockets and all of the bombs wouldn't do much more than slightly annoy the Moon, if the Moon were capable of being annoyed. It's not. We would make a few extra craters in an already incredibly cratered object. We'd also bankrupt the economy of the Earth and pollute our planet horribly in the process. However, there is one thing...

We're going to need some technology we don't quite have yet, but it's reasonable to assume we'll get there. It all starts with a mobile solar panel factory.

This factory produces solar panels from Moon rocks. This is not easy, but the materials are there. Silicon is extracted from the rocks, along with the trace elements required, and turned into solar panels. The panels are built along a ditch that runs along the equator. We dig the ditch to get the materials for the panels. This ditch is important.

Eventually, we have solar panels that cover the entire equator, and a big long ditch. What are we going to do with a ditch?

We're going to build a really big railgun. A railgun is a device that accelerates projectiles on a track, like a levitating train, but much faster.

To do this, we're going to need electromagnets. There's plenty of iron on the Moon, and luckily we have a lot of electricity, thanks to all of those solar panels! Railguns allow us to launch projectiles faster than the escape velocity of the Moon, and with no atmosphere to stop them, these projectiles are leaving the lunar system.

Where do we send them? Well, if we want them vaporized, the sun is the obvious choice. But, we want them out of the solar system, so we have to fire them at another star. Let's use Alpha Centauri—it is the closest star system. Our railgun is going to have to be really amazing to get the projectiles going fast enough, but I believe it can be done. We have an incredibly long path and no wind resistance at all. We're going to need to get these going to around 50 km/s to ensure they leave the solar system altogether—we can't have these things just orbiting in the Oort cloud!

So, we start launching lumps of the Moon toward Alpha Centauri. We're going to need serious mining equipment to slowly dig into the Moon to make projectiles. Luckily, the Moon is mostly solid, so if we plan this properly, we can erode the mass of the Moon in such a way as to make it relatively porous but still support the railgun.

All of that launching will eventually change the orbit of the Moon. This part could be really tricky, but if we're really careful with our mining and the timing of our launches, we should be able to remove huge portions of the mass of the Moon, not mess with its orbit too much, and keep the railgun going. And the best part of all of this—we get to make the Moon much lighter.

Eventually, something has to give. The railgun can no longer operate. Those solar panels are no longer needed. And that's good, because we're going to re-purpose them...as solar sails.

The final chunks of the Swiss cheese Moon are cut up (working from the outside in) and attached to solar sails. This may require some help from Earth in the form of rockets, sails (if we can't use the materials from the panels themselves), and additional workforce, but we're dealing with a lot less mass now, so we should be able to assist. In addition, we might use Earth-based lasers to assist the solar sails along.

Eventually, we've destroyed the Moon. There are several holes in this plan, most notably the fact that the center of the Moon is still quite hot and the difficulty in keeping something like this running for millions of years (the Moon is really big). But the math checks out: You can convert stored solar energy into kinetic energy and use the power of the sun to slowly eject pieces of the Moon into interstellar space. You might even use geothermal energy (lunothermal?) to power the railgun and cool the center of the Moon at the same time, ensuring that you've eventually got solid rock and metal to launch into space.

Oh, there's one more problem: Alpha Centauri has been bombarded with chunks of the Moon for ages. Our aim will never be perfect here, so lots of these will end up orbiting the stars, running into planets, crashing into one another, and creating a great deal of mayhem. If there's any life there, chances are we've caused it all kinds of trouble.

And, if that life is intelligent, they may just build a lunar death launcher of their own.

What are some creative ways to destroy planet Earth?

Don't worry about your safety; I'm just a curious physics student.

The method I used to destroy the Moon could be used to destroy the Earth, but it would be significantly more difficult. Between our enormous gravity, our molten core, and our thick atmosphere, Earth is exceedingly difficult to destroy.

But, many other options won't really "destroy" Earth. Hit it with an asteroid, and it will re-aggregate under its own gravity. Sure, hitting it with a neutron star or throwing it into a black hole would work, but how would we make that happen?

Even moving Earth into the sun is infeasible. How exactly does one move a planet?

One piece at a time, that's how.

So, if you want to destroy the Earth, get busy launching chunks of it into the sun as fast as you can. It'll take a billion years, maybe, but with a little persistence, you too can be a destroyer of worlds, just like me.

If every person on Earth were allowed to kill one other person, how many people would be left alive?

Each individual chooses another to drop dead. Everyone chooses his/her "target" before anyone dies, so people can be targeted more than once.

I think it's safe to assume celebrities/politicians would each be "hit" more than once. *(Editor's note: "Take my word for it…")* But how much so? How many people do you think would be left alive?

Imagine there were only three people in the world, A, B, and C. Also, imagine an equal likelihood of choosing someone for death. What would happen then?

A could choose B or C (50:50)
B could choose A or C (50:50)
C could choose A or B (50:50

The possible outcomes (all equally likely are):

BAA - C lives
BAB - C lives
BCA - Everyone Dies™
BCB - A lives
CAA - B lives
CAB - Everyone Dies™
CCA - B lives
CCB - A lives

Only 25 percent of the time does everyone die. In 75 percent of the cases, there is a survivor. That's one in four for eradication. The average number of people that live is 6/8: 0.75.

If there are four people, only nine of the 81 possible outcomes results in eradication. That's one in nine. The average number of people that live is around 2.8.

For 10 people, only around 0.04% of the scenarios are eradications. And the average number of survivors is around 3.47.

As the number increases, this average number of survivors levels off at 1/e = 36.8%.

Scale up to seven billion people, and the odds that everyone dies is essentially zero. The number of survivors is around 4.424 billion people. That's a lot of survivors.

But, this only works if the odds of selection are equivalent.

In the real world, some celebrities would receive additional selections. This would actually increase the total survivorship—in general people would probably select celebrities more often than not, if for no reason other than they can't think of someone they know they'd really like dead, and they don't know the names of people they don't know!

Oh, and then there are babies, along with adults with mental disabilities—they can't "choose" a victim.

Finally, we have the question of "blindness". Do people get to find out about your vote? If not, people might be more likely to select someone they don't like. But if your vote becomes public, people would probably not choose someone with family that lives nearby—again, more likelihood of celebrities (particularly very nasty celebrities like serial killers and the like) being chosen.

So, how many survive? With blind selection, I'll say five billion. But, with revealed selection, I think it's very nearly seven billion. If people thought their vote would be public, they would regularly choose "bad" people, and the number of them that would come to mind would be pretty small.

What kind of apocalypse would be most harmful to human society?

Hmmm...I was under the impression that "apocalypse" meant "the end of the world". If this is the case, then all apocalypses are created equal (destructed equal?)

But, in terms of the ways that humans could be seriously affected, there are lots of bad options:

1. Asteroid impact
2. Horrible disease
3. War
4. Aliens
5. Cataclysmic volcanic eruption

The worst of these for human society would probably be disease. The others would end things for us more quickly—with disease, the destruction and death would be slower, allowing society to crumble into anarchy.

Personally, I'm hoping for asteroid impact. If we've got to go, let's make it painless and quick.

(Editor's note: (Okay, a shameless attempt to make myself relevant): apocalypse actually translates to "unveiling", "uncovering", "revealing")

What would the world be like if the land area and ocean areas of Earth were switched?

Let's assume that mountain ranges would become ocean trenches the same depth as their current height, and vice versa. Assume that humans developed in 'Africa' which is now where the middle & southern Atlantic Ocean is currently.

I'm going to say it again…I sound like a broken record.

You've just killed every human on the planet.

Betcha can't guess why?

Salt.

There's a lot less water than there used to be, but there's just as much salt as there ever was. The salinity of the oceans is going to increase massively. Even if the salt is on land, it will wash into the oceans, just as it has done here on Earth.

Your new, smaller oceans will be much saltier than our current ones.

This will kill almost all of the photosynthetic creatures that live in the ocean. They will no longer manufacture oxygen.

Their tiny corpses will rot in the blazing sun, consumed by heterotrophic bacteria. The bacteria will consume the precious oxygen we have remaining in their mindless quest to eat all that algae.

Then, those bacteria will die, unable to find enough food to support their massive numbers. They too will be consumed, and this cycle will be repeated, over and over, until the oxygen we need to live is all but gone.

Nearly all oxygen-breathing life vanishes from our world, including us.

Something like this appears to have actually happened, by the way. It's known as an anoxic event. While these events weren't caused by too much salt in the ocean, the end effects would have been similar.

Everyone Dies™.

What would happen if two planets collided at 1 mm/s?

You can ignore the collision. The 1 mm/s is meaningless compared to the gravitational forces between them.

Imagine Earth and Mars "colliding" in this way. To simplify, let's just set them touching one another.

The gravitational force between the two bodies would be 2.67×10^{24} N. This would be enough to accelerate Earth at around $0.44\ m/s^2$. Mars would experience an acceleration of $4.16 m/s^2$

Those are some pretty sizable accelerations—in fact, the force Earth exerts on the surface of Mars is greater than Mars's surface gravity. The materials these planets are made of would be no match for that kind of acceleration.

To put it into simple terms, it would be like placing a building with the mass of Mars on the ground somewhere...the ground simply can't support that kind of weight. Mars would be crushed into the Earth's surface. It would pulverize the crust of the Earth at the point of contact, and bore down through the mantle. All the while the forces between the two bodies would grow as the planets moved closer together.

All of this smashing together causes enormous seismic disturbances on both worlds. The Earth's crust fractures and buckles, allowing magma to spew out around the Earth. In addition, as these planets come together, they lose energy—this lost energy is converted primarily to heat. There's more than enough energy here to melt both planets. Mars continues boring down toward the core of the Earth, all the while melting and breaking apart.

In short order, the two worlds begin to coalesce. After a long while, the crust of our new super planet begins to cool enough to solidify. Mars is no more, and Earth is significantly larger and heavier. Everyone Dies™.

What will our species do when the sun eventually destroys the Earth?

Flee or perish.

If we stay, we're cooked. If we leave, parts of the solar system not currently very nice to live in will be improved temporarily. We could stay within the solar system for a time.

But, we have hundreds of millions of years to prepare for this, and to me there are two good options for humans:

1. Colonize planets in other solar systems. Lots and lots of redundancy for the human species is good for our long-term survival.

2. Modify our bodies to be much more tolerant of less-than-hospitable environments. There is no reason why we cannot incorporate technology to make essentially robotic bodies for ourselves. These bodies could tolerate some environments that are currently off limits for people, including airless worlds or even empty space.

If we progress in these two areas, we should be "extinction proof" for a long time.

(Editor's note: I have $5 that says "Not gonna happen"...)

What would it take to move Venus closer to Earth, making Earth/Venus a twin planet?

An intense desire to kill everyone and everything on Earth. Warning: snarks ahead.

You're still here? You're dedicated. I appreciate that.

So, you still want to do this. Here's how you might pull it off.

We're going to borrow some ideas from: How can we destroy the Moon? (see page 33)

1. You're going to need to acquire an intense desire to kill everyone and everything on Earth. I know I mentioned that already, but it's really important. I'll wait.

2. …

3. …

4. OK, great.

5. Venus is REALLY hot. We can't just build stuff there. So, we're going to have to take the long road—destroy Mercury and Mars first.

6. You might think that ramming asteroids into Venus to speed it up is the way to go. But the mass of asteroids is pretty small—in fact, it's too small to speed up Venus enough to get it out to 1 AU before the sun dies. I want to see this happen, so let's get busy.

7. The mass of Mercury and Mars combined, though, is around 15% of the mass of Venus. That should be enough to "nudge" Venus out to 1 AU in under a billion years.

8. Build yourself two solar-powered railguns, one on Mercury and one on Mars. Launch chunks of those two planets 24/7 such that they impact Venus and increase its orbital velocity.

9. Continue until Mercury and Mars are no more.

10. By amazing coincidence, Mercury + Venus + Mars = almost exactly the same mass as the Earth. I know, right?

11. OK, so now we've got a planet that's almost the exact same mass as Earth, and thanks to our railguns, we've got it in Earth's orbit in under a billion years. Not too shabby.

12. Congratulations—you've successfully moved Venus into Earth's orbit. You may go about your day.

You're still here? Oh, I know why…I can hear you saying it: "Hey wait," you say, "I only read this far because of my intense desire to kill everyone on Earth. All I've done is destroy two planets and changed a third one to a farther-from-the-sun-planet-but-even-hotter-now-thanks-to-trillions-of-impacts-from-space. I thought I got to kill all life on Earth, too. What gives?"

Oh, don't you worry…you'll get what you came for.

You see, the Earth isn't moving at a constant speed around its orbit all the time. Its orbit is elliptical, so it travels faster at some times than others. It also speeds up and slows down ever so slightly thanks to tugs from Jupiter, Saturn and the other planets (yeah, we have two fewer now, but whatever). This means that even if you put Venus on the opposite side of the sun as the Earth, they'd slowly get closer together (either Earth would approach Venus or Venus would approach Earth…it won't matter).

Once the orbits are unbalanced like that, Venus and Earth will start to pull one another together.

This doesn't end well.

Eventually, Venus and Earth will probably collide. This kills everyone and blows Venus and Earth into largish chunks which eventually come together as a very large rocky planet (a "super Earth").

During this process, the new planet is completely molten (Venus was probably still molten from the horrible impacts you inflicted on it, but now Earth gets to be molten, too). If they do not collide, then the orbits of both Venus and Earth become chaotic, possibly throwing one or both of them into the sun or out of the solar system.

And, of course, Everyone Dies™.

What would happen if there were one decillion people on Earth?

10^{33} people? Oh my…

The land area of Earth is around 5.1×10^{14} square meters. This means that on every square meter of land we will need to fit around 2×10^{18} people. That's two billion billions.

Luckily, we can stack people. Using the simple approach of freezing everyone solid, humans are surprisingly easy to stack. It's better this way….believe me. There's not nearly enough oxygen for us all anyway, as you'll see.

So, let's say we can stack three people in a square meter, and that stack is say 33 cm tall. That's 9 people in a cubic meter. So our towers will be around 2.2×10^{17} meters tall.

How far is that?

Well, we've got a stack of frozen humans that goes more than twice the distance from here to Sirius, over eight light years away.

Wait. Did I say "stack"?

I meant stacks. Five hundred trillion of them.

If you prefer, you could also think of this as a ball of humans significantly larger than the sun.

Oh, the humanity!

There isn't enough matter in our solar system to make this many people, but if we did, they'd form a super massive star or a black hole.

That's a much nicer thought than trillions of stacks of frozen carcasses.

What would happen if a comet made of ice about one-tenth Earth-mass were to hit the Earth?

It would depend on the angle of impact. If this dwarf planet (it's too big to be a comet) lightly grazes us, the effect would be *relatively* small. Humanity might survive this, though the climate of our planet would be radically damaged and there would be mass extinctions.

On the other hand, if this object impacted directly, it would most likely eradicate life on Earth. It might actually blow the Earth into pieces, though these would most likely coalesce into a planet again.

So, Everyone (might) Die™.

If every person on Earth turned into a baby, and you had to raise all of them, how many babies do you think would live to adulthood under your care?

Note: You would not turn into a baby, and you would have to raise all the babies you can.

30. Warning - my answer gets a bit dark in places.

I'm the father of two daughters who are now thankfully out of the infant stage. I managed (with an enormous amount of work on the part of my wife) to not kill them accidentally. :)

That said, I think I could keep around 30 alive. This is, by the way, the most awful possible scenario for my life that I can envision, but here's my thinking. I would:

1. Go to a hospital in an urban area that has a river. I'm going with Detroit because I live near there.

2. Put eight male babies and 22 female babies in 30 bassinets in the nursery. Aim for maximum genetic diversity, if at all possible.

3. Go through the keys of all of the pants lying around in the hospital, find the keys to the largest SUV I can find, and find it in the staff lot.

4. Drive it to the grocery store. Fill the entire vehicle with every can of powdered formula I can find. Also, get some large jugs and all the baby bottles in the store. Get thousands of nutrition bars—these are for me. Oh, and get some earplugs.

5. Drive back to the hospital.

6. Make whole jugs of formula at a time. Pour it into the bottles. At first the water supply should work, but before long we'll be stuck with river water. I'm going to have to feed babies nearly around the clock, so getting a system going is crucial. Those earplugs are life and death for my sanity right now.

7. Repeat steps three through six until the babies need food, then move on to baby food and/or cereal. If these babies are newborn, I'm going to be

an absolute wreck of a human for the better part of six months. Forget diapers—just throw poop out the window. Seriously.

8. Once the babies are mobile, this goes from really hard to nightmarish. At first, I can keep them in the nursery of the hospital, but sooner or later, these little ones are going to need to get outside. Finding a park with a fence around it is going to be crucial.

9. Keep this up throughout childhood. These kids are going to have a less than ideal diet, thanks to the constant stream of dried, powdered, and canned food. But, by the time they are teenagers, I should be able to do some farming (maybe we should have started that back at step three…)

10. Ideally, these 30 people should be able to reproduce enough to repopulate the Earth! The genetic diversity will be really low, but it may be viable. This is why I want eight males and 22 females—the goal long term will be to produce the maximum number of children from various parents. This sounds awfully clinical, but the male babies need to try to impregnate multiple females. Fewer than eight and we might not have enough diversity. More than that, and we're probably lowering the number of babies that can be born. Please don't be offended by this—it sounds offensive to me, but remember, this is the end of humanity we're talking about. Sacrifices will have to be made. At the end of their reproductive lives, our 30 humans might be able to make 150 or more humans with a broad spectrum of humanity's genetics.

As if step 10 wasn't dark enough, it gets *much* worse.

There are going to be times when I'm going to need to walk out of the baby room. Maybe I just need to use the bathroom. Maybe I need some air. No matter what, there's gonna be dead babies everywhere, so this will be absolutely awful.

Society will cease to exist. At *best*, I might be able to produce 30 adults who are not completely damaged by the experience of having me as their uber-parent. But did I have time to teach them to read and write? What about math and science? Not only that, but all world languages but English are lost. Opera, dance, basketball, romance novels…I know nothing of these things. They are gone.

So, that's it—the most horrific dystopian future I've ever heard of. I'm going to go have nightmares now. Thanks.

What would happen if the sun lost its light, but still emitted heat?

No photosynthesis!

Without it, we and our fellow air breathers would deplete the oxygen levels to below breathable levels in relatively short order.

We would all die, and our decomposition would further deplete oxygen levels.

All of the plants would die, too, and their decomposition would take oxygen levels to nearly zero.

Mother Nature would do the rest, slowly wearing away the evidence that we had ever existed.

However, life would not cease to exist.

Chemosynthetic bacteria at deep ocean vents and within caves and the like would merrily digest minerals, blithely oblivious to the destruction around them.

Everyone Dies™.

What would happen to the Earth if it stopped for a second and started spinning the other way around?

The Earth has a rotational kinetic energy of around 2.1×10^{29} J. If we stopped rotating and rotated in the other direction, that energy would have to go somewhere.

It takes about 1×10^{8} J to launch a kilogram into orbit, which means we now have enough energy to launch 2.1×10^{21} kg into orbit.

This is roughly equivalent to the mass of water.

ALL of the water. On Earth.

The energy released won't launch all of the water into orbit. It will, however, launch much of the surface water and much of the surface land high into the air.

It will rip the land apart, opening enormous fissures through which molten rock will emerge.

That molten rock will also be flung high into the air, and then rain down starting cataclysmic fires around the planet.

Put another way, this is the same amount of energy as around one million Chicxulub impacts.

It's likely that all life on Earth would be eradicated, though it is possible that some "extremophiles" might survive.

Once again, Everyone Dies™.

How much devastation would a 150 lb rock—going as close as possible to the speed of light without going at the speed of light—do to Earth?

So, the rock has already lost its speed and mass from entering the atmosphere. It impacts going close to the speed of light and weighs 150 pounds. Ignoring if it is even possible or not, how much destruction would happen? (or at least how much energy will the rock have?)

Tiny changes in speed make a huge difference here when we're talking about speeds near the speed of light. As a result, "close as possible" to the speed of light makes this question quite tricky.

But, to give you a few ideas, let's look at some numbers.

First, I'm going to call this a 68kg rock. Metric system!

Let's look at various speeds and compare:

1. Less than 1% of the speed of light. For these speeds, the relativistic effects are small, and the explosions are, too (relatively speaking...we'll be getting into some huge numbers momentarily). At 1% of c, the energy released is equivalent to around 4.87 Hiroshima bombs. Devastation is nearly 100% over an area of several miles, but overall the Earth is unaffected.

2. From 1% to 99% of the speed of light. For these speeds, the relativistic effects start small but build to be pretty significant. At 99% of c, the energy released is equivalent to nearly 9 GT of TNT, or around 600,000 Hiroshima bombs. It is still, however, just a tiny fraction of the amount of energy released by the Chicxulub impact that wiped out the dinosaurs. Whatever you hit at these speeds is instantly obliterated, and regional extinctions are likely the nearer you get to 99% of c, but life survives.

3. Adding 9's after the decimal point. For these speeds, the relativistic effects get enormous, making our 68kg rock behave more like a huge asteroid. At around 0.99999999994c, we have the equivalent of Chicxulub. The planet is devastated, and the Earth is thrown into a hellish climate nightmare of endless darkness, planet-wide fires, and mass extinctions. Life still survives, however. At 0.999999999999996c, we have the equivalent of a

billion Hiroshima bombs, or around 126 Chicxulubs (best unit ever, btw). This probably wipes out life on Earth, or at the very least multicellular life.

4. Obliterating the Earth. The binding energy of the Earth is around 2×10^{32}J. To get our explosion to that energy, we'll need to accelerate our 68kg rock to 0.99999999999999999999999999953c.

   Just calculating this number pressed the limits of my calculating power—we're so close to the speed of light that we've increased the mass of our rock by a factor of nearly 33 trillion. That gives it an effective mass nearly the same as Deimos, a Moon with a radius of over six kilometers.

So, as you can see, when you say close to the speed of light, the degree of closeness makes a big difference!

Need I say it? Everyone Dies™.

What are the top ten reasons to become a math teacher?

The Art of Teaching Math and Science;

"Mathematics, rightly viewed, possesses not only truth, but supreme beauty, a beauty cold and austere, like that of sculpture." — Bertrand Russell.

1. Teaching math is awesome and makes kids smarter and better prepared for life.

10. Alternate number systems help prepare kids for computer programming.

(Editor's note: Bwahahahahaha! This caught me flat-footed…)

How long would we last if the Earth's rotation accelerated by one degree per day every day?

Earth spins 360° per day. At this rate, a year from now the Earth will be spinning at 720° per day, or twice as fast. Keep going, and in less than 24 years, the equator will be spinning so fast that anything sitting on it will be flung into space.

But long before then… Everyone Dies™.

Why? Plate tectonics.

As the Earth spun faster and faster, it would bulge at the equator, slowly becoming more and more oblate. That would be, as the geologists like to say, "very bad".

The African plate especially, and other plates around the equator, would be flexed quite a lot by this. Cracks would appear, magma would ooze out, great earthquakes and tsunamis would rock the landscape. More and more water would "fall" to the equator, shifting the weight of the oceans and causing further tectonic problems. Continents would shift and slide at speeds never before seen, wreaking havoc as they moved.

Oh, and the weather: as the planet spun faster and faster, winds would swirl faster and faster as they were dragged along with it. This would lead to Jupiter-like winds of hundreds of miles per hour on a calm day, knocking over everything from buildings to trees and eroding land masses in a deluge of flying debris. The Earth would be sandblasted smooth by its own abrasive wind.

And yeah, then it would start throwing chunks of itself into orbit.

Chalk this one up to a nightmare I'd rather die than see…nothing like watching your planet rip itself to pieces while you watch.

Which is the most beautiful Moon in our solar system?

I am giving my answer in verse.

Beautiful Moon -

Oh, worlds of such diversity
how can one rate your strange beauty?
Is Io's molten surface fair?
Or Triton made from frozen air?

Speckled face of rugged Callisto?
Or Glorious Titan, in clouds aglow?

Nay, not one of these fine places
Shines with more and sweeter faces
Than Luna, our precious time-worn Moon
Sweet at night and faint at noon.

But ever present in our skies
Light of Galileo's eyes
Without which our sure demise
Thank you, Theia, for our prize

-David Consiglio

Hypothetical Scenario: If Jupiter took the place of Earth and Earth became a Moon of Jupiter, could we survive?

Fascinating question!

I'm going out on a limb a bit here, because there are so many variables I'll have certainly forgotten one. But, I'm going to say "Yes!"

I considered a number of parameters:

1. Temperature
2. Radiation
3. Stability of orbit
4. Weather
5. Increase in asteroid impact
6. Gravitational heating

Let's take them in order:

Temperature: It's going to get colder here, but I don't think cold enough to kill us. We'll be slightly closer to the sun during part of our orbit, which will warm us up, but the distance reduction is trivial. The problem is the shadow. Suppose we orbit at the same distance Europa currently orbits Jupiter. Our period around Jupiter is 85 hours, of which we could spend as much as maybe 15 hours in the shadow of Jupiter. That's something like a 20% reduction in insolation, which is going to make things a bit chillier. Normally I'd think it might be enough to freeze us solid, but I've got some good news later on which will help us a bit in that department.

Radiation: Going near Jupiter is dumb. REALLY dumb. Jupiter spits out an obscene amount of radiation, enough to fry our innards like an egg in relatively short order. But in your scenario, we've got a secret weapon: the magnetosphere. The same shield that protects us from the sun's radiation should protect us from the worst of Jupiter's as well. But there's more energy going into the atmosphere, which means a bit more heat when it's all over. That's actually good news. There's also probably a bit more cancer, which is not so good. But I don't think the radiation will be sufficient to kill us.

Stability of orbit: This one is really tough. If we are orbiting far away from Jupiter, it's possible that the sun's considerable pull on us at apoapsis is enough to perturb our orbit into an elliptical one which eventually becomes unstable. Planets as close to the sun as we are currently don't have big Moons (our Moon is by far the largest). But on the other hand, Jupiter is really massive, so it should be able to lock us in. If we're close, this is more likely to be true, but we encounter another problem: Tidal locking. The major Moons of Jupiter are tidally locked, which means one face points toward Jupiter at all times. Our Moon is tidally locked to Earth for the same reason, which is that the Moon is quite close to a much more massive planet. Europa is a little less than twice as far from Jupiter as the Moon is from the Earth, which suggests that if Earth orbited Jupiter, we'd be tidally locked to the giant planet. But this introduces all sorts of weird scenarios for the amount of light that the Earth would receive from the sun, which affects the...

Weather: If Earth were tidally locked into an orbit that was in the plane of the Jupiter/Sun orbit, then the face that pointed toward Jupiter would never receive direct overhead sunlight—it would be eclipsed every time that would normally happen. It wouldn't even get very much indirect sunlight, as it would be facing away from the sun when it was on the sunward side of Jupiter. Even if the orbit was out of the plane somewhat, Jupiter is so large there would be lots of eclipses. The result is that the Jupiter facing side would get a lot of radiation and not a lot of sun. On the other hand, the side facing away from Jupiter would have something like 21 hours of indirect sun, 21 hours of direct sun, 21 hours of indirect sun, and then 21 hours of darkness. This is certainly survivable from our standpoint, but what would it do to the weather? It's very hard to say, but I think it's safe to assume that the weather would be decidedly different. Overall it's going to be cooler on the planet (due to those eclipses), but how would heat transfer work from the "Hot" side to the "cold" side? It may not be efficient enough to distribute the heat in 20 hours, which would mean one side of the planet would be significantly warmer than the other at all times. This is "not good", as it would tend to dry out. The water would preferentially evaporate from the "hot" side and condense (probably as snow and ice) on the "cold" side. This leaves a large ice cap on the Jupiter facing side, and a dry, hot desert on the away side. However, there are areas near the boundary where it's probably pretty nice.

Increase in asteroid impacts: Jupiter really pulls in asteroids, and now that it's not way out in the solar system, those asteroids are feeling a double tug: Jupiter and the sun are now pulling them in the same direction. As a result, the asteroids are coming. Lots of them. Jupiter is going to absorb the lion's share of the impacts, but Earth will not be spared. But is this survivable? Probably. Even with Jupiter and the sun pulling together instead of apart, space is still really big. Most of those asteroids are going to miss, and even if we get hit by a big one, we'll know it's coming and be able to prepare. Now if it's REALLY big (say Ceres and Vesta), we're toast, but those monsters aren't just going to fly toward the sun—it will take ages and ages to tug them out of their orbits. So at the very least, we should have some time.

Gravitational heating: Being close to Jupiter is dumb (I feel like I said this already), but not just because of the radiation. Jupiter also kneads its Moons like a ball of dough. And, like that dough, the moon's warm up. Earth would become significantly more volcanic as a result of its proximity to Jupiter. This is probably our savior, though. All those volcanoes are going to spew an awful lot of carbon dioxide into the air. Normally that would be bad, but we just put up a HUGE sunscreen in our way every solar days, so we're going to need to keep some extra heat. Losing some of our solar radiation puts us in between current Earth and current Mars in terms of solar irradiance, which means all that CO_2 is a benefit, not a disaster. There will be problems, not the least of which is the sulfur dioxide produced by the volcanoes, which causes acid rain and actually reflects sunlight, but again, Earth has experienced periods of much greater volcanism without killing off everything. We're adaptable!

So, the verdict is in: It won't be easy, and it's possible we get flung into space or crushed by an asteroid impact, but there's a reasonable chance that we could survive.

Extra fun stuff: Jupiter has crazy lightning. We'd easily be able to see it from the Jupiter facing side during eclipses.

Jupiter would look BIG to us (41 times larger than the Moon), which would have it taking up a giant chunk of the sky. It just occurred to me that the Jupiter facing side of Earth would get a lot of reflected light from the surface of Jupiter. The amount of energy here is not trivial (as it is with the Moon), and it's good for us: it warms the colder side of our planet.

We'd be able to see Io, Callisto, and Ganymede, as well as the rings and some of the smaller Moons. They would be perpetually doing a crazy dance in our sky.

We would NOT be able to hold on to our Moon. That orbit is definitely unstable in this situation, so the Moon is lost. Hopefully it's not lost by being smashed into us. :)

We would be able to see volcanoes on Io, maybe even without binoculars.

The aurora would be almost constant, thanks to Jupiter's radiation. They might extend all over the planet.

Radio communication would be much trickier, thanks to all that noise. Radio astronomy would be practically impossible.

Rocket launches and a space elevator might be significantly easier, for a couple of reasons. First, our tidally locked situation means there's one spot on Earth pointed directly at Jupiter. That point is a fair bit easier to launch from, thanks to the pull of Jupiter's gravity. How much easier? Gravity would actually be about 3% weaker on the surface at that spot (and nearly 3% stronger on the other side!) if we were orbiting at the distance Europa does now. In addition, the farther you got from Earth and the closer you got to Jupiter, the weaker that gravity would become. There's a Lagrange point (a point of equal gravity) in between Earth and Jupiter, which at Europa's distance is only about 22,000 miles up! That's closer than geosynchronous satellites are currently (about 26,000 miles). It's not much, but it might make that small amount of difference that makes the space elevator possible. It might also be enough to slowly rip off our atmosphere, so don't get too excited. :)

If the Earth really were flat, would we really "go over the edge" if we came to its end?

This is actually a really interesting question. Let's suppose the Earth was a disc, like a CD, and we lived on "top" of that disc. Right off the bat, we would not feel gravity as we do today. Depending on our position on the disc, gravity would pull us to the left or right, not up or down. This is actually somewhat interesting in that there is only one place on the disc where gravity would seem "normal": On the edge. If we were walking on the edge, gravity would pull us down and we would not be able to tell we were on a disc just by feeling the force of gravity.

But walking to either face of the disc would radically alter that situation—we would fall "down" toward the center of the disc.

So, if we were on the edge of the disc, walking off either side would cause us to fall, almost like falling off of a huge cliff. The other way around, walking off of a face and onto the edge, is impossible.

You're somehow stuck on the Moon with unlimited supply of food, water, air, and a shovel—how do you let the people of Earth know something is up?

The Moon is mostly covered with a fine powder. You could easily make patterns in this using your shovel. It would take many years, but you could write huge letters or symbols on the surface of the Moon that would be visible to the naked eye on Earth!

May I humbly suggest a crop circle, just to mess with humanity for ages to come.

Or, perhaps "HLEP". Just because how funny would it be to spell help wrong?

If Alpha Centauri were to die in a giant supernova explosion, what would it look like from Earth?

None of the three stars in the Alpha Centauri system are large enough to go supernova. This, as you will see, is a good thing!

However, suppose one of them was. The most recent nearby supernova took place in the Crab Nebula—and it was visible during daylight hours. It is also more than 6,000 light years away.

Alpha Centauri is only about four light years away. Since it is 1,500 times closer it would be about 1,500*1,500 = 2.25 million times brighter if it had the same intrinsic brightness as the Crab Nebula supernova.

How bright is that? At least as bright as the sun, and significantly more dangerous. Supernovae emit powerful radiation, and at this distance Earth would be bathed in a nasty barrage of charged particles and gamma radiation. The charged particles would be partially deflected by the Earth's magnetic field, though the intensity might be enough to strip away our atmosphere altogether.

However, the gamma rays would pass unhindered to the surface, wiping out most (if not all) life on the planet. It would be like getting a million x-rays in a day.

It would be beautiful...briefly. The sky would be ablaze with a new and powerful light. Then, we would be dead.

Everyone Dies™.

Could we make a landing on Jupiter's Moon Io?

It depends on what you mean by "we". If you mean humans, well, things might not go so well.

First, you'll need to make the transit from Earth to Jupiter. This will probably take a few years, so sit back and enjoy the radiation, I mean ride. Ride. Outside of the Earth's protective magnetosphere, you'll be regularly irradiated by low-level solar and cosmic radiation. This probably kills you, but if not, there are always coronal mass ejections to worry about. These are giant explosions that come out of the sun and bathe everything in their wake in extra deadly radiation. Awesome, no?

OK, but let's say we've shielded the craft from radiation. Next up: Jupiter space. This is a region even more packed with sweet, delicious, radiation-y goodness. You'll love the experience of having your DNA pelted into submission by the ever present glow of Jupiter's emissions.

Not a fun place to live and all of the equipment we send there has to be specially designed to withstand the nonstop pounding Jupiter puts out. But again, it is technically possible to shield against the radiation.

Even if we live this long, we've still got Io to contend with. A tectonically active lava ball so close to Jupiter that it has tides of *rock* higher than the tides of water here on Earth. So "land" isn't really a thing so much as you might think.

Sure, there's a solid surface over much of the Moon, but it is in a constant state of change, perpetually cracking, melting, sliding, and shaking. Not a great place to build a base.

But, after all of that, could we land on the surface of Io? Yes. Could we survive that trip, and the trip home, too? Probably not.

With current tech it would be extremely difficult to survive the trip there, much less Jupiter's radiation and Io's nasty surface.

Perhaps one day, thousands of years hence, we might put a real human on Io. I suspect that for the short and mid-term future, it will be flybys and probes and robots only.

Would people survive the onslaught if asteroids were to destroy every major city in the world?

In a word: No.*

In a few words: Almost certainly not.

At first glance, you might think that there would be areas unaffected by asteroids. And actually, that would have been true several thousand years ago.

We just have a LOT of cities now.

Roughly one half of the Earth's population lives in cities. Now, you said "major cities", which is a bit vague, but we have a lot of those, too. Let's say we wanted to JUST destroy New York City (which is not the largest city in the world, but is definitely "major").

Now, we need to discuss exactly what "destroy" means. The atomic bomb the US dropped on Hiroshima didn't completely destroy the city, but it came pretty close. So, if we use "wipes out the buildings and kills almost everyone" as a metric, then we can say that since the Hiroshima bomb was equivalent to about 15 kT TNT then we'd need maybe 100 kT to 'destroy' NYC. This is an asteroid in the somewhere in the 15—130 meter diameter size, depending on air burst or ground landing. Let's say 100 meters to destroy a major city—lots of assumptions here, but it's as good a number as any.

So, we need a 100 m asteroid to destroy a major city. How many major cities are there? If we assume a major city has 100,000 people in it, then it turns out there are around 4,000 major cities on Earth.

Your mileage may vary if you think 100,000 isn't a major city, but I'm writing this here answer so I'm going with it. :)

Anyway, that's 4,000 impacts of a 100m object, each with a 100 kT yield. That's 400 MT of yield. In all of history, we've blown off about 510 MT in nuclear testing, so one would think that humanity would survive the onslaught.

However, there are two important differences. First, that testing took place over decades—I'm assuming that you are implying that all of the cities on Earth were destroyed simultaneously.

And, second, most of these tests took place underground—our asteroids would be hitting the surface or exploding in the air.

The combination of 4,000 surface impacts would produce an absolutely enormous amount of dust and debris (and maybe even magma!) that would be ejected into the atmosphere, blocking out the sun for years.

In addition, every forest on Earth would be set ablaze. The object that killed the dinosaurs set North America on fire all by itself—all of these smaller impacts would burn the Earth. This would produce an enormous amount of soot, further blocking out the sun.

But, believe it or not, all of THAT might be survivable, though certainly not for most people.

The nail in the coffin is that the oxygen content in our atmosphere would drop dramatically. Burning all of that biomass all at once would consume significant amounts of our atmospheric oxygen. And with the sun blocked out by all the dust, photosynthesis would essentially halt. Plants would no longer be replacing the oxygen gas the fires were using. Then, bacteria would begin to consume all of the dead animals and people, using even more oxygen. I think you can see where this is going: almost everything that needs oxygen to live would not have enough to get by.

*Now for that asterisk. There is a whole wide universe out there! If we knew this was coming with enough lead time, we could build a self-sufficient colony on Mars, the Moon, or some other body. This would allow us to ride out the destruction in a safe location, and potentially repopulate the Earth after millennia of recovery. But, there are a nearly infinite number of potential disasters that could occur during those millennia that would doom humanity anyway. This colony would have to be huge, multiply redundant, and well populated with diverse and intelligent people. But, barring that, we're toast.

Everyone Fries™.

(Editor's note: Bwahahahaha See what I did there?)

If there was a speck of dust, 100x the heat of the sun, what would it look like, and what would happen if I let it roam around in my room?

OK, let's take the energy of the sun, multiply by 100, and put it in your room. What could possibly go wrong?

According to this, the sun puts out about 3.8 x 10^{26} J every second. That is to say, the sun is essentially a 380,000,000,000,000,000,000,000,000 Watt light bulb.

We're multiplying by 100, so we're going to build a 38,000,000,000,000,000,000,000,000,000 Watt bulb.

Oh, and it's going to be as small as a speck of dust. That doesn't matter, really...let's just go with a light bulb.

(Editor's note: Just attempting to paragraph format those numbers was a nightmare- I gave up.)

Let's say we install this light bulb in your room, but it's turned off. It looks like an ordinary bulb. You go to the wall, look at the bulb, and flip the switch.

This is a really poor choice on your part. I'm not judging, but you've just killed everyone on Earth.

But first, what does it look like?

It doesn't look like anything. The light and heat energy travel out from your light bulb just like they would any light bulb, at the speed of light. And that energy travels the distance from the bulb to your eyes in just a tiny fraction of a second.

But, unlike an ordinary bulb, they rip the air molecules apart on their way to you. This emits additional nasty radiation, but you have nothing to worry about! That's because when the light gets to you, it vaporizes you from front to back. Imagine making a house of cards and then smashing into that house of cards with a 747. It's kinda like that. The chemical pigments in your retina change shape with those first incoming photons, just like they are supposed to ... in the picosecond before your eyeballs, and then your entire body, turn into plasma.

The wall of energy does not stop in your room. It proceeds to vaporize your house, your city, your country, and your planet, in about 0.043 seconds.

Oh, but we're not done yet.

We've created an energy source similar to that of a supergiant star. As that energy radiates outward in a sphere, it eventually passes the orbit of Mars. The distance from Earth to Mars varies, but on average it's about 200 million km.

At this point, the energy from our light bulb has grown to a sphere with a radius (r) of 2×10^{11} meters. The surface area of that sphere is $4\pi r^2$ which is about 5×10^{23} square meters. That's an enormous sphere.

Surely, at this distance, we're safe, right?

Each square meter of Mars receives $1/(5 \times 10^{23})$ of that energy, which comes out to around 58,000 W/m^2.

The sun on Mars only provides about 589 W/m^2, so if we were standing on Mars, our light bulb would look about 100 times brighter than the sun normally does there, or about 42 times brighter than it does here on Earth.

It's also about six times brighter than the sun looks on Mercury, a place which is pretty hot!

Your light bulb melts the surface of Mars.

The compounds on the surface begin to break down, releasing atomic hydrogen and oxygen from the planet.

Mars slowly erodes away over many years to a shadow of its former self. The energy here might be sufficient to melt the entire planet.

Venus and Mercury are probably destroyed altogether.

But, it's not a total disaster.

Titan ends up with a new sun that spits out about 2100 W/m^2. This is a bit higher than what Earth gets from the sun (Titan gets about 15 W/m^2 from the sun, so we can ignore that).

But, Titan is smaller than the Earth, so a little extra heat will probably do it good. Titan is the new tropical destination for all of the exiled humans.

Everyone Dies™.

What would happen if someone built a shrink ray and stole the Moon?

Assume that the person stealing it is from Earth.

Gru is a cartoon with an impossible facial configuration, but let's say he's normal man sized. I'm normal man sized, and my hand is about 10cm across. The Moon just about fits into his hand, (I have kids-I have watched the movie. A lot.) so, let's give it a diameter of 10cm.

The newly compressed Moon has an enormous density, so high that it compresses protons and electrons into neutrons, and then further forces neutrons to occupy the same space. The Pauli Exclusion Principle states that no two identical fermions (which include neutrons) can exist with exactly the same quantum state. This means some of the neutrons in our compressed Moon must exist in higher energy states than they would normally.

They're not terribly fond of this.

I previously concluded that the compression was so great that a region within the compressed Moon would have sufficient gravity to overwhelm this pressure and condense into a black hole. I'm still holding onto a shred of that position, as you'll see later. But I'm almost certainly wrong. I'll defend my original answer (just a little) below, but assuming I am wrong, here's what actually happens:

Most likely, when we stop compressing the neutrons, they leave their high-energy state and fall into a lower energy state. This liberates enormous a mounts of energy which causes a massive explosion. The energy of explosion would be within a few orders of magnitude of the mass-energy of the Moon. Using 1×10^{20} kg (which is about three orders of magnitude less than the total mass of the Moon), we get an energy of explosion of 9×10^{36} J. This is an approximation (these things are difficult to know with certainty), but as you'll see, it doesn't much matter.

This is about the same as the energy output of the sun puts out…in nearly 1,000 years.

This isn't just an explosion. This is the end of the world.

The Moon is a mere 240,000 miles from the Earth, and it is now releasing in a few moments the equivalent of 1,000 years of solar radiation. Even if we're off by a factor of 10 (or even 100), this rips the atmosphere from the Earth before vaporizing much of our planet. Everyone Dies™. I was right about one thing: Gru most definitely misses the dance recital. However, he does not disappoint the "leetle gurls", as they are now vapor.

OK, now for the asterisk—is making a black hole still possible here? Our compressed Moon is now a billion times denser than known neutron stars. Only one force (that we know of) is capable of compressing objects to this density: Gravity. But, Gru is clearly not using gravity to shrink the Moon. He's using some other force. This is a cartoon, mind you, so we don't really know what that force is. What we do know is that the properties of matter at these densities are very poorly understood. There is no question that we are forming some kind of Degenerate matter, but exactly what type is not clear. The densities are (in my opinion, anyway) far too high for this to be normal neutron-degenerate matter and, as such, there is still the outside chance that our compressed Moon begins to convert fermions into bosons. Bosons can exist in infinite numbers in the same quantum state, so these newly-formed bosons could, just maybe, allow for the formation of a singularity in the center of our compressed Moon.

Finally, as I was writing this, another thing occurred to me: Gru carries the Moon back to his ship after he shrinks it. Gru is able to move the Moon with his hands, which means it's very lightweight. Getting rid of this mass by converting it into energy annihilates the Earth, killing everyone. If we just "disappear" the mass, that also goes badly for us (though not nearly as badly). We simply don't have a Moon, which messes with our orbit. However, since Gru puts the Moon back, it's probably OK in the end.

What are the biggest misconceptions about you?

"Dave is the Destroyer of Worlds."

That's true, but I'm a complex fellow. I have feelings and needs, the same as any man. I like long walks on the beach, smooth jazz, and curling up with a good book in front of a fireplace. My perfect day is spent on a sailboat eating amazing seafood and drinking cocktails that someone else has made for me.

I'm a sensitive individual.

No one <sniff> ever asks me how I feel. It's just "Dave, what would happen if Earth were hit with a fish the size of Pluto?" and "Dave, can the Earth be crushed into a black hole using powerful magnets?"

It gets a little lonely, destroying worlds.

Now, if you'll excuse me, I'm going to play "Everybody Hurts" by REM and have a good cry.

But, then I think…"You know…this song title is awfully close to my trademark…maybe we could play this really loud while the asteroid hurtles closer and closer…"

And, then I feel better. Who needs sailing? Screw jazz!

Back to work!

What would happen if the Earth was hit by a fish the size of Pluto?

I see what you did there…

but, I have a job to do.

Everyone Dies™.

Pluto has an enormous mass (around 1.3 x 10^{22} kg, or around 1/450th of the mass of the Earth). If this fish swam through the interplanetary abyss and ran into Earth, it would crack our planet like an egg.

What would happen if all of Earth's oxygen turned into ozone?

Ozone is a much more powerful oxidizing agent than oxygen.

In English, nearly everything that could be oxidized would be oxidized, and that right quick.

Nearly all of the ozone would be consumed in short order, forming carbon dioxide and other oxides. There would also be no oxygen for humans to breathe. 20% ozone would oxidize your proteins, rip your lungs to shreds, and then burn your corpse in a planet-wide conflagration.

So, yes, Everyone Dies™.

Environmental science: what would happen if carbon dioxide were completely removed from the Earth's atmosphere?

It's known that too much CO_2 is bad. What would be the other extreme of too little CO_2?

Oh goody, oh goody, oh goody…

Everyone Dies™.

Without CO_2, every plant on Earth would perish in very short order. Photosynthesis would screech to a halt. The heterotrophs of the world, mostly bacteria and animals, would digest the plant material on land and in the sea, would consume all that dead plant material.

That digestion would make CO_2, but not nearly quickly enough to save any plants who might have survived the original die-off.

But, what this digestion *would* do is use up all of our oxygen. In mere days, the oxygen levels would be too low to support humans. Shortly thereafter, nearly every oxygen-consuming organism on Earth would be dead.

This would leave just anaerobic life, and there is lots of it. Well, there was lots of it. Now it would absolutely overrun the planet. Without oxygen (which is toxic to anaerobes), these bacteria would occupy every niche on Earth, digesting organic material and multiplying at an alarming rate.

The by-product of their metabolism is often methane (CH_4) which is a potent greenhouse gas, many times more potent than CO_2. The planet would warm quickly, roasting the anaerobes and boiling the oceans.

Most anaerobes couldn't survive on this hellish world, but some could—extremophiles. They'd be in their glory, living all over the Earth in large quantities.

Our atmosphere, now loaded with CH_4, would be a home to all sorts of life—and that life might repeat a trick it pulled off once, billions of years ago. One day, many years after you've killed everyone, an anaerobic extremophile might evolve photosynthesis. If it did, it would slowly consume the CO_2 and make oxygen. That oxygen would first react with the methane, making more CO_2. Eventually, though, almost all of the CO_2 in our atmosphere would be replaced with oxygen.

And, then maybe, just maybe, oxygen-breathing life might evolve. Again.

What are the qualities you would want for a virus meant to eradicate all humans?

You have to create a virus that will make all Homo sapiens extinct, what two (you can choose up to five if you want) characteristics would you choose for your virus to fulfill this duty?

P.S. This is just for fun.

P.P.S. I am not a biological terrorist.

1. High transmissibility.

2. High lethality after a lengthy, symptom-free incubation.

Consider two diseases: Influenza and HIV. Influenza has the first of our two criteria totally covered. It's incredibly easy to get the flu. Someone sneezes, or you drink out another person's cup, and voilà. You're sick. But influenza doesn't kill most of the people who get it. It does kill, but not enough to meet your criteria.

Now, compare that with HIV. HIV has the second criteria totally covered. Untreated, HIV has a near 100% fatality rate, and there's a good chance you don't know you have it for quite a while. But it's hard to get. You have to have unprotected sex with an infected person, and even then you might not get the disease. You could also share needles. But sexual activity and IV drug use leave out large swathes of the population, including children, many older folks, and the celibate.

OK, let's put them together.

A new virus arrives on the scene. It's readily transmitted. It proliferates in saliva, nasal mucus, blood, semen, vaginal fluids, etc. It's as easy to get as a cold.

This is no ordinary virus. It can survive on metal surfaces for months. It can lie dormant on a dirty fork or a child's toy nearly indefinitely.

When you get this disease, you suffer no symptoms. In the meantime, the virus is infecting your immune cells. It is slowly lowering your white count until you are defenseless against all nature of pathogens.

If this disease existed, and we were infected without our knowledge, it would kill nearly everyone within a decade. We'd try very hard to treat it, but look at how long HIV ravaged the population before we had any real treatment for it. If nearly everyone on Earth had been infected with HIV in 1982, there would be almost no one left on Earth to research treatment.

But, to kill everyone, this disease would have to have no natural immunity in the population. This is the toughest part—humans are a diverse group. It's likely that among the seven billion humans on Earth, a few of them are immune. If there are enough of them, they could get together and repopulate the Earth with their immune offspring.

If their numbers were too small, though, they might die out.

What would be the global effects of receiving a message from Proxima Centauri?

Assume there are Aliens that are exactly as advanced as we are.

It's a message similar to the Arecibo message and scientists can decipher it. What would they do next?

When is the public informed? What would be the (long and short term) effects of it?

Twitter would BLOW UP!

All joking aside, we wouldn't have to translate it to know what it means: "INTELLIGENT LIFE RIGHT HERE!!!"

If that's the case, the short and long term effects are profound. First, we have the existential revelation that we are not alone in the universe. Indeed, there are other sentient beings practically on our doorstep. Second, the ball is in our court. They may not know we exist (though they probably do), but either way it's our turn to respond. How we respond may dictate the future of humanity.

Now, our obvious first step is to translate the message. We need to make sure it doesn't say something to the effect of "We are the Borg. Resistance is futile. You will be assimilated."

Chances are it will say something like "Here we are. We know math. Do you know math?"

So, if the message is friendly, step two is to get the best nerds around to write a message back in their language. Hopefully this says something like "Hello. We also know math. Care to establish a conversation?"

Then, we wait for at least nine years.

The travel time for the signal is 4.3 years or so, followed by a 4.3 year return trip. Also, add some time for them to craft their response.

So, what do we do with those nine years? We build ships to go to Alpha Centauri. Ships that can get humans there in relatively short order (i.e. before they die). That's an enormous undertaking which will span far longer than nine years of work, but now is the time to do it. In fact, we might even want to delay our response a bit to start the building process first. We want to be able to go to *them,* not the other way around.

We may eventually want them here. That said, it's far easier to control the overall situation if you send a few humans there. Worst case scenario is that the aliens eat those people for lunch. But, at least they are years away *there* rather than right here.

Step three, hopefully, is the free exchange of information. There is so much to learn from them, about them, and with them that it staggers the imagination.

You are tasked with writing a fictional story about one or more Quorans. Who would you pick, and what would the story be about?

Let your creativity flow!

This story is about Habib Fanny.

(Editor's Note: THIS was one of the reasons that this man impresses me!)

I do a lot of walking.

Tonight, my walk is in the rain. People scurry past as I walk under the elevated tracks. The last train of the night clatters past. Car horns blare in the distance. The smell of cheap beer and deep fried food wafts into the street from a nearby dive bar. I don't hurry—I'm just a guy walking in the rain.

Nothing to see here.

My employer demands detail and precision in every job, but particularly one this high profile. That's why he chose me.

I'm an assassin, and when I come for you, you die.

It's not like what you think, being an assassin. People think it's all killing, but I hardly do any killing. Mostly I walk.

You see, anyone can kill—that's easy. But, in my line of work, it has to be done perfectly. That means no witnesses, no evidence, no loose strings.

So, I move to a new town. I walk the streets at night in the places the target likes to go. I learn where the he eats, what bars he goes to, what kind of music he likes, who he's married to and who he's screwing on the side.

I do a lot of walking.

Marco is the mayor of this town. Normally, I don't hit politicians because the heat puts me out of commission for too long. But my employer wanted me specifically for this job. He didn't say why; he never says why. He never says much of anything. Mostly, it's just a name, maybe a photo, and a sheet of paper with a phone number and that curious phrase:

Everyone Dies™.

I don't ask—I just do my job. But, I got a feeling why someone might want Marco killed. He's pushing hard against organized crime. It's not because he wants to clean up the city, mind you. No, he's pushing hard because he wants to be in charge of the crime in this town. He wants the money funneling through city hall, not the local mob bosses.

Maybe one of those bosses pulled the trigger on this hit. Maybe he hired my employer to make Marco disappear.

Maybe not.

It doesn't matter. I peer into the window of the Mexican joint where Marco is still stuffing his face. Marco can stuff stupid amounts of refried beans into his oversized face hole. Little does he know that he's eating his last meal.

So, why tonight? Tonight is perfect. For starters, there's the rain: rain makes everything better for me. It muffles any sound this dumb mook might make. I'll garrote him so quickly he won't have time to talk. But, sometimes guys flail, and Marco's got a hundred pounds on me easy. If he kicks over a trash can or falls into something noisy, people will just think it's the rain, if they hear it at all.

The rain's good for more than muffling the noise, though. It washes away footprints, blood, everything.

And, people don't poke around into other people's business in the rain. People stay home and watch TV when it rains like they're supposed to.

Marco is still stuffing his face, so I start walking up the street again. When you hold still people notice you, but when you're walking you're invisible. Up the street, around the corner, down the street, across the street, back to the cantina.

Marco is on his way out now. Of course, he doesn't pay, and the last thing I see before I put my head down is the owner's disgusted face. He's paying protection money to Marco and the local mob boss; it's bleeding him dry and this jerk doesn't even pay for his nachos. A guy like that deserves to die if you ask me.

But, no one asks me. They just give me a job, and I do the job.

Marco walks out of the restaurant and down the street. He's with his two enormous bodyguards as usual. These two are unusually large, even as bodyguards go, but they're no good at their job. I've been watching them work for weeks, and they let the boss out of their sight all the time.

I walk past, turn the corner, and rush to take my position. My speed alarms no one—it's raining.

The bodyguards don't worry me because they're walking Marco to an apartment where one of his girlfriends lives. He likes to visit Angela on Thursdays…never did figure out why he liked her on Thursday. Doesn't matter now. She'll never see him. Right on cue, the bodyguards stop at the front entrance to the alley while Marco walks right in, past the dumpster and the back entrance to the liquor store to the stairs that will take him to Angela's.

I'm under those stairs, soaked to the bone and a little out of breath, but calm. It's dark back here, and Marco's eyes still haven't adjusted. So, when I say "Hey" in a soft voice, Marco doesn't notice my faint French accent or my ebon skin. He can't see or hear much of anything. That's ok—I just want him to pause, which he dutifully does.

This is my favorite part.

I move swiftly around behind him, wrap the wire around his neck, and crush his windpipe. All those years in medical school did one thing for me—I know how to destroy a man's larynx in a second.

Marco falls to his knees in panic, lets out a soft gurgle, and then falls face first onto the wet pavement of the alley.

I keep the pressure up on the garrote for a few minutes, just to make sure. Marco has been very accommodating in the way he fell and I only have to drag his gelatinous bulk a few feet to the alcove beneath the steps where I'd been waiting for him. I pull out my phone and take a photo. The flash goes off, but on such a perfect night a faint shimmer of lightning covers my action.

By Horus, I love a night like this.

I walk casually out the back entrance of the alley and into the night. I hope I never return to this town. I probably never will.

I text the number my employer gave me with the job. I attach the photo.

He always sends the same reply: "A+".

Never figured that out, either. Probably never will. It doesn't matter—the deposit notification appears on my phone a few minutes later. By then, I'm more than half a mile from Marco's rapidly cooling body. He won't be found for another two hours and, by that time, I'll have walked all the way to the train station and onto the next assignment.

I do a lot of walking.

(Editor's note: This is gold. Pure gold.)

How expensive would the average Everyone Dies™ scenario be?

Considering the costs to develop new technologies and create/transport parts.

OK, I decided to take this one on, not in hypothetical units we use in our hypothetical business, but in actual US dollars.

The only way we could manage this scenario with today's technology would be to push a large asteroid into a collision course with Earth. Nuclear weapons won't do—there's no realistic way to get them all fired. Even if we did, a few might survive, and we just can't have that. Diseases never wipe out everyone because natural immunities will almost certainly exist somewhere in the population.

No, there really is only one way to be sure: hit Earth with something huge.

So, what do we use? And what would that cost?

For this devastation, we're going to use 433 Eros. The irony is not lost on this nerd—I *love* this stuff.

This bad boy has a mass of around 6.7×10^{15} kg, which is around five times the mass of Deimos, and roughly the same mass as the Chicxulub Impactor. That's the rock that wiped out the dinosaurs.

OK, so we've got our space rock. This guy gets within around .15AU at the closest. We're going to need to modify its orbit to get it to intersect the Earth's orbit. This is rather complex due to the eccentricity of Eros's orbit as well as the uncertainty of when this is being done, but I'm going to estimate about a 1000 m/s Δv requirement in order to move Eros into a collision course with Earth.

I estimate that, using many ion engines like the ones on the Dawn spacecraft, we will need about 7×10^{14} kg of propellant to "correct" Eros's orbit.

The fuel for Dawn was pure Xenon, which cost NASA roughly $250/kg to purchase. At that price, the fuel for just the orbital correction would be $175 quadrillion (with a q). The cost of hauling that much fuel into place via conventional means (from Earth) would be around $10,000/kg, or $7 quintillion. The cost of hauling the engines to Eros would add to the cost.

The total amount of Xenon in Earth's atmosphere is nowhere close to what we'd need, requiring us to use less effective Krypton or sourcing our Xenon from other worlds. Yikes!

So, all in all, I'm going to estimate the total cost of this mission at roughly $10 quintillion. That's 1×10^{19} dollars!

What does your 100,000 years of gross world product buy you?

Each Dawn engine provides around 90 mN of thrust, but we'll need over 500,000,000 such engines to do the job. This will provide roughly 46.5 million Newtons of force, enough to accelerate Eros at the snails-leave-us-in-the-dust rate of just 0.000000006940299 m/s^2 (that's 6.9 nm/s^2).

The closest Eros gets to Earth is around 2.2×10^{10} meters. Now, obviously, everything is moving around constantly, but just to give you an idea, to travel that distance starting from rest and accelerating at our glacial pace, you'll need around 3.6 billion seconds, or roughly 114 years to get Eros to Earth. I'd be surprised if we could pull this off in under 500 years, what with all of the launches and assembly.

But, when Eros gets here, it'll be going at an alarming rate. If we time this right, we should be able to get an impact speed in the 25,000 m/s range, faster than Chicxulub (which we'll need because Eros is a bit lighter).

What happens next will be the stuff of legend.

Eros will land in north central Asia, carefully aimed to do the most damage. Why there? Well, for starters, the earthquake this generates will likely wipe out the populations of Europe and Asia, where the bulk of humanity lives. What's more, the massive boreal forests will all burn, as will the boreal forests of North America, thanks to sub-orbital magma and debris cascading around the planet.

Several billion people do not survive the 24 hour period following the impact.

In the years to come, choking dust and soot from planet-wide fires halt photosynthesis, choke rivers and lakes, and poison the oceans. No land animal larger than a rat survives, and nothing in the sea larger than a crocodile makes it through.

Humanity is toast. Everyone Dies™.

Dave Consiglio: When Everyone Dies™, do you die too?

Nope.

"Everyone" refers to the unlucky souls terminated by the customers of Consiglio Devastations. If you hire us to devastate a population, "Everyone" in that population gets devastated or you get your money back. Guaranteed.

If you choose to devastate the Earth, then Consiglio Devastations reserves the right to save various members of the team, as it sees fit, for the purposes of carrying out the devastation.

As protection against accidental devastation, I have created numerous backups of both my consciousness as well as cloned bodies. Rest assured, the devastations will continue, even if the original Dave Consiglio does not.

Does Dave Consiglio get annoyed by Quorans who ask why nobody died?

On many of Dave's answers that are more political or simply just not about everyone dying, I can find several comments that point out that nobody died. Especially on more serious topics, does Dave get annoyed by this?

Very occasionally.

Everyone Dies™ has made me very famous (notorious?) here on Quora, and for that I'm very grateful. It's the "Jeremy" to my Pearl Jam, the "Freebird" to my Skynyrd, the "Stairway to Heaven" to my Led Zeppelin. You have to love it, even if it gets overplayed a bit.

But, if someone shouts out "PLAY FREEBIRD" in the middle of your mom's eulogy, yeah that's kind of annoying.

Occasionally, I get the "Nobody died!!?!1??" as a comment to questions like "How can I be a better physics teacher?" and it doesn't really make a lot of sense.

Even then, though, I think it comes from a good place. I have had nothing in my life to prepare me for the kind of attention I've gotten in the last six months, and I'm still trying to navigate it.

In the end, I try to think of it as a good thing. Every "So does everyone die™??" commented in an answer is a person who might learn something by reading my work. In my core I'm a teacher—if they're reading, they're learning, and if they're learning, then my time has been well spent.

Everyone Learns™.

What would happen if the center of the Earth suddenly turned into osmium?

Let's start with the replacement of the inner core.

OK, the inner core has a radius of around 1.22×10^6 meters and a density of around 12.8 g/cm^3. That's a fair bit denser than the normal density of the materials that make up the core (iron and nickel mostly) thanks to the pressure exerted by Earth's significant gravity. Since the density of such a mixture would be around 8.4 g/cm^3, we can guess that a roughly 50% increase in density will take place based on the pressure introduced by the weight of the outer core, mantle, and crust pressing down on the inner core.

If that holds true, then osmium (around 22.6 g/cm^3) will be around 33.9 g/cm^3. This is roughly 2.65 times the density of our inner core, and so our inner core will be 2.65 times more massive.

Amazingly, the mass of the inner core represents only about 1.7% of the mass of the Earth. This mass increases by a factor of 2.65, but as a result the mass of the Earth only jumps by about 4%. Everything on Earth is about 4% heavier…which does almost nothing. A few buildings fall down. Everyone feels a bit more tired. Maybe some tall trees collapse. That's about it. I know, surprisingly boring!

If we scale up to the inner and outer core, things get more interesting.

Now, we're talking about roughly 1/3rd of the mass of the Earth, but much of that outer core has a lower density than the inner core. This means that the Earth increases in mass something closer to 75-100%, and with that extra mass comes a near doubling of gravity.

Thanks to that extra gravity, Everyone Dies™. Why? You will very quickly become exhausted just moving around under these conditions. Walking around is now like walking around while carrying your twin. Rough. Old people collapse on the spot, and even the fittest among us are incapable of moving for any length of time. Our blood becomes much heavier, too, and our hearts labor to pump it around. Our massive weight makes breathing difficult, and we slowly asphyxiate and suffer from heart failure.

Oh, also we probably have no magnetic field. Not that it matters—the radiation from space is the least of our worries. Most of us are stuck to the floor in the kitchen or bathroom, our legs and wrists broken and writhing (slightly…remember how heavy we are) in pain. The pain is brief, mercifully.

Oh, also buildings mostly fall down as do forests. And tunnels and pipes collapse. And roads and bridges. Lots of other stuff, too. Houses fall on top of us. Great fun.

Let's ramp it up all the way to "everything under the crust". At this point we've multiplied the mass of the Earth by something like five (the mantle is relatively low density) and everything on planet Earth is now five times as heavy as it once was. I'm nearly half a ton now, and the weight of my blood rips my arteries in multiple places. I bleed to death almost instantly, unable to scream due to the massive weight of my chest.

Oh, and the new Earth pulls the Moon down on top of it. Not that we care.

Everyone Dies™.

What would happen if the speed of sound and the speed of light changed places?

Everyone Dies™.

Your biology depends on chemical reactions which take place very quickly. Electronic transitions enable very rapid transitions in the shapes of molecules. These shape changes enable our bodies to metabolize, see, synthesize DNA, and a number of other critically important functions.

The disruption of these pathways disrupts homeostasis and Everyone Dies™ in short order.

The sounds of their screams travel at the speed of light.

What would happen if all the water in Europa's ocean was teleported to Earth?

The water would be distributed evenly over all of Earth's surface.

There are around 3×10^{18} cubic meters of water on Europa.

If Earth were uniformly flat, this would create a spherical shell whose thickness can be calculated using the formula:

$$V = \frac{4}{3}\pi\left(R^3 - r^3\right)$$

In this case, "r" is the radius of the Earth. Since the Earth isn't a perfect sphere, this will be off by a bit, but using the standard 6,371,000 meters as the radius, and solving for R, we get:

$$R = \sqrt[3]{\frac{3\left(V + \frac{4}{3}\pi r^3\right)}{4\pi}}$$

Plugging in our values gives us R = 6,376,876 meters. Subtracting the radius of the Earth, we get: 5,876 meters.

How high is that?

Well, the highest inhabited place on Earth is La Rinconada at 5,486 meters.

These folks only have 390 meters of water over their heads.

The height of the water would actually be a little higher than that because we're assuming a uniform spherical shell. However, we ignored the land. That land will push the water up even higher.

Simply put, if Europa's water was teleported to Earth, it would flood all but the very highest mountains.

Everyone Dies™.

What would happen if all drinking water was polluted with barium?

We'd have to treat it with sulfate in order to precipitate it out.

Everyone Lives™!

Oh wait, we don't know it and so we don't treat it?

The approximate LD_{50} for barium is around 200 mg/kg/day.

The solubility of barium nitrate is around 105,000 mg/kg, so if the water is polluted anywhere near the saturation point of barium nitrate we all get a huge dose of barium before we're aware of it.

Everyone Dies™.

How much mass would an object have to contain to have a Schwarzschild radius the size of the sun?

The Schwarzschild Radius of a non-rotating black hole is given by:

$$R_s = \frac{2MG}{c^2}$$

Solving this for mass, we get:

$$M = \frac{R_s c^2}{2G}$$

Plugging in the radius of the sun (6.957×10^8 meters) and the values for c and g, we get:

4.69×10^{35} kg

This is about 235,000 times the mass of the sun, or roughly the mass of a globular star cluster.

Heavy!

If a black hole was heading towards our sun, what would happen?

If a black hole were headed toward the sun, it would gain speed due to the mutual gravitational attraction between it and the sun. The sun would also accelerate toward it, as would all of the planets. But, if we're talking about a stellar mass black hole, the deflection in our orbit would be small enough that we wouldn't all die right away (unless of course the black hole gets very close to us).

But worry not—we're doomed.

When the black hole comes close to the sun, its gravity begins to rip the sun apart. At first, this results in a significant increase in brightness. Then, as matter from the sun begins to swirl around the black hole, it is accelerated to enormous speeds and it begins to emit high energy X-rays and gamma rays. This radiation will bathe the Earth in its deadly glow, but even that might not kill us all.

And, that's because you've aimed the black hole directly at the sun. So, our sun gets pulled more or less straight in. Some material will swirl around for a bit, irradiating us, but most of it will be absorbed into the maw of the beast straight away.

What remains for Earth is total darkness.

I know not whether we end in fire (gamma radiation) or ice (cooling to single digits Kelvin thanks to the complete lack of a sun), but both are just as nice.

Everyone Dies™.

What would happen if an object (let's say a man) on Earth suddenly gained more mass than Earth itself? What would happen to gravity and spacetime?

There are a couple of ways to approach this, but in both of them Everyone Dies™.

I'm also going to answer this as if *I* were that man, because reasons.

First, let's ignore the fact that I'm made of something like neutronium.

The presence of my mighty mass this close to the Earth would create an enormous gravitational disturbance. I would literally suck the land, the air, the seas into my mighty bulk as I tunneled at enormous speed toward the core of the planet. Earth and I would grapple for mass, ripping each other to pieces before coalescing into a planet with twice the mass of the Earth. Oh, and we pull the Moon down on top of us, making planet Davearthmoon even bigger. Oh, and I put my name first. Don't like it? Tough. I HAVE THE MASS OF PLANETS.

Second, let's stop ignoring the fact that I'm made of something like neutronium.

I am made of neutronium! The next time I do something awesome in the presence of my wife, I'm totally saying this, by the way. Luckily I will get that chance very soon as I am a nearly-perpetual fountain of awesomeness. Sure I sound arrogant…but I am made of neutronium! What have *you* done lately?

Unfortunately for Earth, neutronium is only stable under gravitational fields far stronger than I can exert. My gorgeous physique (I'm just gonna lay it on thick because I am *literally* thick with neutrons) lasts mere microseconds before exploding with the force of trillions of megatons of TNT. The explosion rips the world to shreds before ripping the Moon to shreds. The enormous cloud of rubble probably recongeals to form a planet…probably.

No matter how you slice it, Everyone Dies™ and I am totally responsible for it.

I am made of neutronium and it feels fantastic!

What would happen if Earth's core and the Moon changed places?

Everyone Dies™.

The Earth's core is either 760 miles in radius (inner) or 2,160 miles in radius (inner and outer) depending on which one you want to go with.

Inner Core: The Moon, suddenly transported into a space far too small for it, would expand with extraordinary force. The shock wave would travel at the speed of sound in the mantle (something like three miles per second) in all directions toward the surface of the Earth.

Since the Earth has a diameter of around 7,917 miles, the surface is something like 3,000 miles from the surface of the now-subterranean Moon. This means that the shockwave will take around 16 minutes to travel to the surface.

When it does, it rips the crust apart and throws it (and everything on it) clear into space.

Oh, and around the shattered remnants of what was once our beautiful Earth there would be an orbiting glowing iron ball. It weighs a lot more than the Moon did. The collisions with the shrapnel you've created (nice job, by the way) slow it down significantly and it spirals into the Earth, colliding with the still molten Earth. The collision further disrupts our planet, sterilizing what has already been sterilized.

Eventually, the new Earth-Moon conglomerate recongeals into a larger, heavier world without a Moon.

Inner and Outer Core: The Moon, with its radius of just over 1,000 miles, leaves a very large gap between itself and the mantle more than a thousand miles above. The Earth collapses in under its own (and the Moon's) gravity. The opposite fate befalls the denizens of Earth—instead of being blown into space, the ground beneath our feet literally falls away and we are sucked down over a thousand miles to our death in the molten nightmare below.

In this case, the orbiting glowing iron ball has far too much inertia to stay in its orbit. In fact, the "Moon" now weighs significantly more than the "Earth", so the orbital configuration will be chaotic to say the least.

Still, though, Everyone Dies™. I can sleep easy.

What would happen if all of the nitrogen on the Earth disappeared?

Everyone Dies™.

The proteins in your body, as well as DNA, all contain lots of nitrogen. These would literally fall to pieces, killing you, and every living being on Earth, instantly.

Even if we just limit ourselves to molecular nitrogen, Everyone Dies™. Atmospheric pressure would instantly decrease by around 80%. This would result in water evaporating very rapidly. Puddles in hot areas would start boiling. Human beings would dehydrate very quickly. All that extra water vapor in the air would act as a greenhouse gas, massively warming the planet. Without atmospheric nitrogen, no more nitrogen would enter the biosphere and all plants and animals would die that way, too.

You just killed everyone in at least four different ways (disintegration, dehydration, roasting, and nutrient deprivation).

Impressive.

What would happen if you compressed each of the following to half their current size and suddenly let go:

1. Earth
2. The Moon
3. Mars?

The following answer is long and math-y. If that's your thing, read on. If not, skip to the break for the summary.

First, we could go with half the diameter, half the surface area or half the volume and still be justified in calling it "size". That said, the answers in all three cases are pretty similar. Note: I had to make numerous approximations in order to get a reasonable answer.

Solids are almost incompressible because in order to get atoms any closer together, you'd have to push their electron clouds into one another. Check out this photo:

This was taken using an electron microscope. What you see there are gold atoms, all packed together in a nice orderly array. Notice the spaces between them? That's where the electron cloud of each atom comes near to another.

Since electrons are all negative, they repel one another. The result is that if you try to squeeze atoms close together, this forces negative electrons closer to other negative electrons. If you let go, the repulsion will push the atoms apart.

This happens all the time. You compress the floor ever so slightly when you walk across it, and the floor rebounds when you return.

But, what you're asking about is *a lot* more compression. To accomplish this, you're going to have to overlap electron clouds by quite a bit. This is going to store lots and lots of electrostatic potential energy in those electron clouds. You're compressing a perfect spring using an enormous amount of stored energy.

How much stored energy?

This is very tough—planets are not uniformly constructed. What's more, this deformation is likely to be somewhat elastic (the rebound is back to normal size and shape) and somewhat inelastic (some materials don't return to their initial conditions).

Let's look at a simpler example:

1. Let's assume elastic compression—not sure how true this is, but the compression will be at least partially elastic. It makes the math doable, though, so I'm going with it.

2. We're going to model this as 2-D compression. This makes the math a lot easier (I'm not as differential equation friendly as I once was...math nerd out there, feel free to go hard core on this one). By ignoring the 3rd dimension, our estimate will be too low.

3. The Young's Modulus of basalt is 60 GPA. Steel is around 200 GPA. Let's model a planet as an object that has a Young's Modulus of around 100 GPA (in between rock and metal).

The energy stored in this scenario is given by:

$$U = \frac{E A_o \Delta L^2}{2 L_o}$$

Here E is Young's Modulus, Ao is the original surface area (the surface are of the Moon or planet), ΔL is the change in length of the 2D object in question, and Lo is the initial length of the object.

Plugging in our numbers for Earth, we get:

$E = 1 \times 10^{11}$ GPa
$A = 5.1 \times 10^{14}$ m^2
Lo = Radius of Earth = 3.2×10^6 m
$\Delta L = 1.6 \times 10^6$ m
$U = 2 \times 10^{31}$ J

This is around 1/10th the gravitational binding energy of the Earth. Since we've estimated too low (we're ignoring the "3D" component of the compression), I'm going to estimate that the stored energy is roughly equivalent to the gravitational binding energy of the object. The gravitational binding energy is the energy holding a celestial body together, and thus our body now has enough stored energy to blow itself into interstellar space.

Compressing a body by half *very roughly* gives it enough energy to literally blow itself apart. The actual answer probably lies somewhere between "Huge portions of the planet are blasted into interplanetary space" and "The entire planet is blown into interplanetary space".

This answer makes sense when you consider another object that undergoes massive compression similar to what you described:

A supernova.

In a supernova, a star ceases to produce enough energy via fusion to hold back the gravitational forces that pull it inward. The resulting explosion blasts enormous amounts of material away from the star and into interstellar space.

What you've asked about is not nearly as energetic, but then the gravitational forces holding a planet together are not nearly as strong. My rough estimate is that it doesn't matter if we're talking about the Moon, Mars, or the Earth—any solid object (and all three of those bodies are 'solid-ish') compressed by 50% will rebound in such a way as to blow itself to smithereens.

Oh, and in the case of Earth, it almost certainly goes without saying, but just to be consistent:

Everyone Dies™.

What would happen if the sun turned into a tennis ball for one second?

I'll assume the sun turns back into a sun after the second is up. If it does, the Earth is unlikely to notice the effects. It would get just a bit cooler, but one second is not long enough to cause and significant climate change or problems for life on Earth. This would be similar to planet-wide night for one second—hardly something to worry about.

But, one has to wonder how the sun is being turned into a tennis ball…

If the matter of the sun were compressed into a tennis ball, we'd have a much different outcome.

The Schwarzschild radius for the sun is around 3,000 meters. Since a tennis ball is significantly smaller than this, in this scenario the sun becomes a black hole.

That planet-wide night lasts longer than a second—this is a forever night.

The planets still orbit the sun as much as they did before, revolving around in near total darkness. With only starlight and Hawking radiation to warm us, the water, people, and atmosphere of our planet freeze solid. Only hydrothermal vents and subterranean bacteria stand a chance.

There's also the possibility that, during the compression, an explosion of sorts takes place, much like a supernova, but in miniature. If that were to happen, Earth would be irradiated with sterilizing radiation prior to freezing solid.

Everyone Dies™.

Of course, it's possible that religion as we know it is wrong, and Serena Williams is really the creator of the universe. I've heard more implausible hypotheses!

Imagine Goddess Serena, holding the sun in her divine grasp.

The good news is that Serena is a benevolent goddess, merciful and forgiving.

The bad news is that she has grown tired of our impetuous nature and she has incredible aim:

At the speed Ms. Williams can fire a tennis ball, there can be only one outcome: Everyone Dies™.

How would you destroy the world using only Spam (the food product)?

Scurvy.

If I made everyone eat nothing but Spam, everyone would die from vitamin deficiencies.

Most noteworthy of these is scurvy. In fact, sailors in days gone by used to get scurvy precisely because they ate a diet devoid of vegetables and fruits (which did not keep well on a ship) and instead ate a lot of salted meat (spam), bread, etc.

In addition, Vitamin A deficiency seems likely, as well as anemia caused by iron deficiency.

Spam is also horribly high in saturated fat and devoid of fiber, a diet that is likely to lead to colorectal cancer. Finally, Spam is dripping with salt, which could lead to high blood pressure and heart disease.

Destroy the world with Spam? Easy. Just make people *eat* it.

Everyone Dies™.

(Editor's note: I could have told you that...)

Why can we not see the stars through the sky during the day?

You know, this is a really good question. I spent a long time writing this, editing, and re-writing. Then I did some research because I wasn't happy with my answer.

Seems like an easy question, no?

OK, first the "standard" answer: Scattering (aka "glare")

When you are on the Earth trying to see space, you can't because you are seeing sunlight that is scattered by the air molecules and dust in its vicinity and then into your eyes. You see a lot of scattered sunlight shining into your eyes from the entire sky, and this scattered light is a lot brighter than starlight. In fact, since short wavelengths are preferentially scattered, you see those colors more than you see the colors associated with longer wavelengths—this is why the sky looks blue.

And, there's the trick—you can still see the stars, their light is simply drowned out by the sheer *amount* of scattered light.

So, why doesn't the reverse happen? The answer appears to be "it does". You just can't see it because it's too dim.

Here's the explanation. The sun is *very* bright. Almost all of its light reaches the Earth unscattered. A small percentage gets scattered, and that's why the sky looks blue and not black. But, a small percentage of sunlight is enormously bright, more than enough to overpower the relatively dim stars.

The light reflected by the Earth is *much* dimmer, mostly because almost all of the sun's light is absorbed by the Earth—Earth isn't very reflective. Much of what is reflected is infrared, which your eyes can't see.

Just like sunlight coming in, most of the reflected light passes straight through the atmosphere unscattered. This is what astronauts see from orbit. A tiny percentage is scattered, but this amount is so dim as to be invisible or nearly invisible to the naked eye. This is made more prominent by the fact that most pictures are taken straight "down" at the Earth, where the scattering is smallest.

You *can* see that scattering of reflected light if you look right at the horizon. There you are only seeing scattered light (because the air is effectively transparent). The horizon looks blue, but not high-intensity blue.

Short answer: the sun's brightness results in so much scattered light that you can't see the stars. The Earth's comparative dimness results in a proportionally smaller amount of scattered light, allowing you to see the ground and the oceans, even in "daylight".

Which astronomical object had the highest percentage probability of causing serious damage to the Earth in recent history?

Well, the objects that hit us have a 100% chance of hitting us. :) So every object that's hit the Earth had the highest probability of causing serious damage. That includes Tunguska, Chelyabinsk meteor, Meteor Crater in AZ (not sure if that's recent enough for you), etc.

But, perhaps you mean what object had the highest percentage of causing serious damage but *didn't* actually hit us.

In that case you're almost certainly interested in 99942 Apophis.

Of all of the Near Earth Objects we've charted, only Apophis has come close enough to us to really get us nervous.

Apophis had an initial likelihood of 2.7% of hitting us—that has now been downgraded to essentially zero. But 2.7% is about the highest scientists have ever seen.

If Apophis hits us, it will cause widespread destruction and loss of life, definitely enough to qualify as "serious damage". But it would not likely be an extinction level event. It would cause a major tsunami if it landed in the ocean, and a very large crater (larger than the one mile Meteor Crater in AZ) if it impacted on land. Anyone within several miles of the impact would be killed instantly, and the corresponding climate disruption would probably kill many more. Nearby forests would be set ablaze and molten rock would be thrown high into the air (and into orbit).

We are very lucky that Apophis is not going to hit us!

Because if it did, well, you know.

What would happen if a super dense cube of volume one cm^3 but mass equal to the Moon were to suddenly appear on Earth's surface?

Let's assume this chunk to be denser than neutron star material (maybe the aliens could somehow manipulate Higgs bosons to make that happen). I am interested in the gravitational and electromagnetic impacts on the Earth and the planetary system and of course on the human beings.

One cubic centimeter is not small enough to make a black hole out of a lunar mass. Instead, compressing the Moon that small would convert all of the matter to neutrons in a very high energy state. The minute you put it on the Earth, it would explode with enough energy to literally rip the planet apart. Please never do this :)

Everyone Dies™.

What if water on Earth was replaced with heavy water?

(Editor's note: This is my thinly veiled attempt to add something meaningful to everything else that has been written. It was this or a dirty limerick…)

Now THIS is some creepy timing—I am currently researching this as we speak!

Here are some conclusions as of yet: I am drawing on the writings of Josh Velson for a lot of my information.

Deuterium oxide has properties that are quite different from light water. it will be more dense (!), have both higher freezing and boiling points, lower surface tension, and a higher viscosity (thickness) .

This will get interesting after the chart!

The scary part, however, is that it will have both a higher heat of vaporization and heat of fusion (chart is from Wikipedia).

Properties[9]

H_2O (Light water)	HDO (Semi-heavy water)	D_2O (Heavy water)	Properties[9]
0.0 °C (32 °F) (273.15 K)	2.04 °C (35.7 °F) (275.19 K)	3.82 °C (38.9 °F) (276.97 K)	Freezing point
100.0 °C (212 °F) (373.15 K)	100.7 °C (213.3 °F) (373.85 K)	101.4 °C (214.5 °F) (374.55 K)	Boiling point
0.9982	1.054	1.1056	Density at STP (g/mL)
3.98 °C[10]		11.6 °C	Temp. of maximum density

H_2O (Light water)	HDO (Semi-heavy water)	D_2O (Heavy water)	Properties[9]
1.0016	1.1248	1.2467	Dynamic viscosity (at 20 °C, mPa•s)
0.07198	0.07193	0.07187	Surface tension (at 25 °C, N/m)
6.00678	6.227	6.132	Heat of fusion (kJ/mol)
40.657		41.521	Heat of vaporisation (kJ/mol)
7.0	7.266 (sometimes "pHD")	7.43 (sometimes "pD")	pH (at 25 °C)
1.33335	1.32844		Refractive index (at 20 °C, 0.5893 μm)[11]

Let's start with the differences between heavy water and the regular 'I-can-drown-you' stuff:

Suppose that we made the change suddenly, at least in geology time terms:

It is going to make the water approx 11% heavier—this will smash anything that doesn't get into shallower waters quickly. Fish, crabs, submarines at their depth limits. I don't think that they have this much of a safety rating (the submarines, not the crabs.)

Ships would suddenly have a hard time staying upright as their buoyancy will have suddenly changed. They may not fall over; but roughly a 9% increase in floating ability is definitely going to throw some things off. Possibly literally. Even though they pop up a little, it will take more fuel to overcome the increased water resistance, both from the increased hull drag and also the drag from the props chopping through the water.

Water skimmers and other things that rely on surface tension of the sea might start having a harder time, and to a lesser extent creatures that float on the water will have a hard time with balance, although I don't like seagulls so those jerks can just tip over.

A huge percentage of the ocean would explosively freeze due to higher freezing points—this instant phase change would not only create a savage shockwave though both air and sea, but it would also release a ton of heat into the atmosphere—especially at the polar regions. Waterspouts, mini-hurricanes, and some pretty scary winds would develop (Ever read *Cat's Cradle*—specifically Ice 9?)

Every single floating ice cap in the world would suddenly have enormous buoyant forces placed on it due to greater mass displacement (even with the increased weight of ice, the water weight would increase even more). The effects are most pronounced in Antarctica. Some ice sheets break apart in an epic release of potential energy, and others "bounce" immediately, causing tidal waves from the distortions. Hard to believe that Roland Emmerich missed this in *The Day After.*

All waves in the world suddenly become less exciting, as viscosity increases. Surfing begins to suck—that is, it does until the tidal waves shows up.

All living creatures in the sea suddenly have a severe pH imbalance. Some might be able to adapt based on their cellular buffering mechanisms, but the majority will not be as lucky...

Because of the increase in heat of evaporation, the world suffers a sudden worldwide drought and the humidity level and precipitation amount drops precipitously (bwahahaha! See what I did there?) Famine on a global scale—both the water and the lack of water kill you.

This is copied directly from Velson: "Replacing the volume of all the world's oceans water with deuterium oxide also affects the mass of the planet and gravity. **According to the National Oceanic and Atmospheric Administration**, there are 1.335 billion cubic kilometers of water in the oceans, about 1.335 x 10^{21} kg; changing all of that to deuterium oxide will add 1.48 x 10^{20} kg of mass to the Earth. Since the Earth's mass is 5.97219 x 10^{24} kg, the total mass of the Earth increases by 0.0025 percent.

Gravity will be correspondingly higher by just that amount.

GPS would stop working, most geostationary satellites would need to be adjusted, and many other space objects would begin to de-orbit.

If the Earth's momentum changes you can count on every single satellite in the sky coming down or going way, way up, due to the change in the number of magnetic lines of force going through the satellite in any given time. This would either increase it's "drag" or decrease it to the point that it might leave orbit. A failure to account for the passing through the magnetic lines of force (each time it caused a slight decrease in momentum) was a major contributor in the premature Skylab re-entry a decade or so ago.

The Moon's orbit would change, with its periapsis and apoapsis coming closer (thanks, Kerbal Space Program!). The period of its orbit decreasing slightly, but fortunately the Earth's additional mass would exactly cancel out the additional gravitational forces between the Earth and the sun. Less than fortunately, the question doesn't specify the angular momentum and linear momentum of the additional mass, so I can't comment on how the Earth's rotation might change, except to note that it will probably cause more tidal waves.

A catastrophic change in the volume expansion—from the change in mass relative to the change in density. Since deuterium oxide has a mass of 20 Daltons and light water has a mass of 18 Daltons, the mass of water in the seas increases by about 11.1 percent. However, density has increased by only 10.6 percent! So, the total volume of the oceans suddenly rises by 0.45 percent. Man, Emmerich has really missed out.

This would be the mother of all eustatic sea level rises. If you expanded all the volume of the ocean by 0.45 percent, assuming a surface area of 360 million square km, it would result in a massive 54.75 feet of water above the current surface of the sea. (OK, it's a guess—but it's a REASONABLE guess). That's 16.69 meters. Expect some inconceivable floods and a lot less real estate. Our beach house on Bald Head is gonna get creamed.

The tectonic plates are going to take a beating—kiss California goodbye. Of course, that would also take out the Kardashians, so it's not a total loss.

The tides become less extreme due to the higher density involved—which will only be a comfort for the survivors left in Denver.

Now, after that little bit of fun, we have got some long-term (bwwhahaha! I kill myself—it's almost as if we HAVE a long-term…) deleterious effects:

The heavy water will begin to show up in our water supply:

- First - desalinated water
- Second - surface water
- Third - resources from precipitation
- Finally – groundwater

Within the first day or so, we'd notice the differences. Deuterium water tastes worse (sweeter, by documentation), probably a little acidic, and will be heavier (!) than expected.

In addition to bloat (because of volume expansion) and a perceptible increase in weight, in the long term deuterium oxide is toxic.

Deuterium oxide toxicity becomes an issue when 25 percent to 75 percent of our body's water is replaced with it. 50% is considered to be the point of no return.

There is some lower blood pressure, but nothing too serious. At least, not in the scale of events.

From animal evidence, however, we know that 25 percent bodily content of deuterated water makes us irreversibly sterile (back to my Kardashians comment—at least it's not ALL bad news).

When our levels gets to above half, the buffers in our physiological systems get wrecked. Read: fatal, mourners please omit flowers.

Plants will probably do better, but the lower surface tension in heavy water and the higher density means that trees will die above a certain height as the maximum height that water can be brought to by surface tension comes way down. The redwoods are going to be the first to go—the clue will be that they die from the tops on down.

Because heavy water makes eukaryotic cell division impossible due to its impact on the mitotic spindle, plant seeds will not germinate when only fed heavy water.

Plants will live but will be unable to grow or regenerate lost cells.

Most multi-cellular eukaryotic plant species are doomed—and guess where Homo Sap gets most of his oxygen?

In addition, because water spontaneously ionizes and reforms, any light water that mixes with deuterium oxide will almost instantly become a partially deuterated water (back to that Ice-9 comment I made earlier).

When the rains fall, they will soon drop larger (due to lower surface tension) and heavier raindrops filled with thoroughly deuterated water. All joking aside, you thought the dents from the last hailstorm were bad...

Within a year or so, most bodies of water will have adapted to the new equilibrium. Nearly all eukaryotic organisms will go extinct within a few years.

Since most heavy industry operates relatively closed cycle with respect to water, process industry won't start failing until significant precipitation changes the properties of the makeup water. Assuming that anyone is left to operate the machinery, that is.

Coolant loops would be affected first, while boilers could hold on at least in the short term..

Industries that need to continue functioning would need to implement light water filtration systems while the rest of the world frantically redesigned all steam systems to use the new properties of water.

On the plus side, the just-in-time inventory system of most plutonium reactors would no longer be needed as procurement problems of deuterium would cease to exist.

And, let's not forget, the REALLY SERIOUS implications of this, mainly:

No more beer, due to (at least in part) yeast fatalities because of an increase in osmotic pressure

and

Squirt guns wouldn't shoot as far.

The water has turned to poison, the earthquakes, tidal waves, rising sea levels, and social upheaval taking most of their brethren, the last humans will survive:

- Drinking bottled water,

- Pumping the last "light" water from deep aquifers, Lake Vladivostok, Lake Baikal, and the Evian bottling factory (Le Bleu gets their water from a relatively shallow aquifer, so they'd be out of the game quickly).

- Desperately running hydroponics facilities with carefully hoarded viable seeds next to heavy water working fluid, nuclear-powered light water refining centers.

From high ground. Which is, of course, the only ground left.

What would happen if you instantaneously moved Uranus so that it was touching Earth?

Then, your Mom would be touching Uranus. What follows is a lot of puns. You've been warned.

OK, we've got ourselves a truly wonderful Everyone Dies™ here.

Uranus is around 14.5 times more massive and around four times larger (by diameter) than Earth. So, if you happen to be lucky enough to be on the side facing Uranus, you're going to see a massive greenish disk occupying a massive chunk of the sky, like the Hulk mooning you RIGHT in front of your face.

Briefly.

Then Earth will be pulled toward Uranus, like the planetary equivalent of a suppository (it gets worse).

Unlike Earth, with its paper-thin atmosphere, Uranus has an atmosphere that's roughly 1/8th of the thickness of the world. This means that Earth is able to sink well into the atmosphere without striking anything "solid". Simple Newtonian physics gives you an acceleration of around 5.6 m/s^2 of Earth toward Uranus. This means that people will stay stuck to the ground (where the acceleration is 9.8 m/s^2 in the opposite direction) and not sucked into Uranus (seriously, these puns just write themselves). Our whole planet, though, will move toward Uranus at ever increasing speed.

The problem with this simple analysis is that the acceleration on the Uranus side will be far greater than that on the far side. The resulting tidal forces will rip Earth apart.

But, for a bit, anyway, Earth will keep it together. This will be an amazing show as Earth travels the 6,000 km or so through the atmosphere toward the mantle. The whole time Earth will be tearing itself to pieces, or digested if you will (you won't), but I think there's a reasonable chance that some people live long enough to watch massive quantities of light gases from Uranus's atmosphere pushed into Earth's atmosphere. That means if you're really lucky, you'll get to asphyxiate from the toxic gases from Uranus (uh huh) while watching the Earth literally shoved into Uranus (I went there twice).

After roughly a day of travel (it's really hard to calculate this—how does one estimate the friction of Uranus? Ok…I'm really sorry…that one *literally* just happened) the tattered remains of Earth hit the mantle. This layer is a plastic, flexible solid (like feces, really) so if any of the Earth survives, it will "impact" here and mix in with the mantle. Denser iron and heavy metals will continue to sink toward the core, where they will eventually settle months to years from now. Lighter materials from the Earth will boil outward and "outgas" into Uranus's atmosphere. (No, seriously, this is the technical term).

On the outside of Uranus, there is now a massive blemish on the cloud tops. The aforementioned outgassing of colored gases, as well as debris from Earth, disrupt the uniform greenness of Uranus, and leave a huge brown eye on the surface of Uranus (had enough yet?)

Parts of Earth that were on the far side of the impact are thrown by the collective rotations of the two bodies into orbit around the larger world, making a giant ring around Uranus (aren't you glad you kept reading?)

In the end, we've totally pounded Uranus (I'm nearly done, really) but due to its flexibility and the relatively slow speed of impact, Uranus doesn't crack, but rebounds into its original enormous round shape…like your mom.

Everyone Dies™—thankfully, as they will no longer have to suffer through the diarrhea of puns this answer issued forth.

THE 'END'.

Afterword

Very few people on this rock can lay claim to being both intelligent and also possessing the talent to convey that intelligence to others.

My first exposure to Dave Consiglio was on a midnight run to St Louis; I was working in the computer security field, so most of my employment was late-night. Or early morning; depends on how one wanted to look at it...

As one could imagine, not a whole lot can be done on the computer or cell phone in the car at 70 mph at one a.m., but I had just installed a Text-to-Voice reader that could let me keep up with the new projects on Quora (a question and answer website that has begin to eat up a substantial amount of my time—it is fantastic!)

The very first item that my program begins to read is Dave Consiglio's *"What Would Happen if You Instantly Moved Uranus so That it was Touching the Earth"*—between Dave's snarky (but very accurate answers) and the very proper British (Sue) voice on my system, I almost put the car in the ditch.

I literally had to stop for almost five minutes, as I was laughing so hard I was crying and simply couldn't see the road. It has been a long time since I have wet my pants (yes, over a month), but I came very, very close this time!

I promptly looked up Mr. Consiglio—turns out that he has a treasure trove of similar answers to some really thought provoking questions!

As a writer myself, I thought that this man needed to be heard beyond the bounds of Quora—everyone will most likely enjoy some part of what he writes.

Seriously, I would have beaten small children with a two-by-four to have had this man a teacher; well, I would have tried, anyway—the little savages always fight back.

Some of the questions were really bizarre (*"What Happens if Earth is Hit by a fish the size of Pluto?"*), but Consiglio makes an effort to work out exactly what would happen should that all-but-impossible guppy decides to take a swipe at 'Terra Very Firma'.

From there, he started to receive all sorts of "End-of-the-World" questions; in the midst of striving for plausible and defensible answers, he coined the buzzword phrase for his newly formed "Planetary Annihilation" company (I suggested Consiglio Devastations: "We Give You the BEST Bang for your Buck!"—he responded with "Consiglio Devastations, Where our deadlines **really** are!") that has become a calling cry for the masses yearning for information—"Everyone Dies™".

Anytime you have an educator that is attempting to inject fun behind REAL learning, then you have just met someone that is going to change the face of the Earth.

Our goal in presenting this is to try to bring some more students into the world of imagination, critical thinking, (no matter how bizarre) and the realization that enjoyment can be active, as well as passive.

I had a ball editing this—I really think that you are going to enjoy it!

Eddie Wetmore
February 1, 2017

About the Authors

Dave Consiglio, Jr.

David Consiglio, Jr. is a high school chemistry and physics teacher, nerd-in-residence, and knower of useless things. In recent years, he's taken to answering silly questions at *quora.com.* He specializes in wantonly destroying worlds in creative and unusual ways. David lives with his wife, two daughters, and two dogs in Michigan.

Eddie Wetmore

Eddie Wetmore is a part time author and full-time inventor. He runs the Blog and Podcast "Creating Riches from Ground Zero" and also writes the "Fifty Sheds of Gray" stories online. He is a contributing author to 'Folk' magazine and has been since its inception. His latest projects include the "Oasis Edible Garden Project", the Duo-view Glasses project, and running top-speed in a desperate attempt to keep up with David Consiglio. He lives at home (sort of) with his wife, is abused by eleven Rottweilers, and pretty much spends his time feeding anything that shows up on his doorstep looking like it could use a meal.

Stay tuned for more mayhem and destructive frivolity
when Happy Endings Volume Two comes out!

Consiglio Devastations, LLC
Where our deadlines REALLY are!

Made in the USA
Middletown, DE
29 December 2018